From the
Agriculture
Course
to Sustainable
Farming

From the Agriculture Course to Sustainable Farming

100 Years of the Biodynamic Movement

RUDI BIND *and* **UELI HURTER**

Floris Books

Translated by Bernard Jarman

First published in German as *Biodynamisch! Geburtsstunde
der biodynamischen Landwirtschaft am Ausgangspunkt der
Ökobewegung* by Verlag am Goetheanum, Dornach in 2023
First published in English by Floris Books, Edinburgh in 2025
© 2023 Sektion für Landwirtschaft
English version © 2025 Floris Books
Rudi Bind and Ueli Hurter have asserted their rights under the
Copyright, Designs and Patent Act 1988 to be identified as the
Authors of this Work
All rights reserved. No part of this book may be reproduced
without prior permission of Floris Books, Edinburgh
www.florisbooks.co.uk

e | Also available as an eBook

British Library CIP data available
ISBN 978-1-78250-942-4

Contents

Introduction

In the face of the climate crisis, biodiversity loss, soil erosion and the pollution of our water resources, the question of how to live a healthy, sustainable life becomes an ever more pressing one. Part of the answer can be found in healthy food grown from healthy soil, and so, biodynamic farmers and gardeners, food processors, traders, researchers and plant breeders have been pioneers of sustainability for the last hundred years. Throughout the world they have been working with farm-dependent livestock systems, with innovative plant-breeding practices and manuring that makes use of biodynamic preparations, and they have developed new forms of land ownership and food chain networks. And all of this is informed by the idea of the farm as a living organism, as an individuality.

Where does this innovative spirit come from? The answer can be found in Rudolf Steiner's Agriculture Course, a series of eight lectures he gave to 130 people in June 1924. In that moment a new agriculture was born and the biodynamic movement began. For many of the participants, the course was a life-changing experience. They dedicated themselves to this new 'art of farming' with an intense enthusiasm that they then passed on to succeeding generations. Through research and practice, understanding of the contents of the course deepened and new ways of working together were found. As a result of these efforts over the last one hundred years, biodynamic agriculture has spread throughout the world.

This short book describes how the Agriculture Course came about and what it sought to accomplish. We trace how the course influenced not only the biodynamic movement but also the wider organic movement and growing ecological awareness in society. Finally, we give examples of initiatives and enterprises that characterise the biodynamic movement today.

We very much hope that this book will serve to inspire farmers and gardeners – those who have been practising biodynamics for years or who are considering converting to it – as well as students, teachers and co-workers in various initiatives all over the world.

1. The Agriculture Course

A crisis in agriculture

In the aftermath of the First World War, agriculture in Europe faced a crisis. Soils had become acidic and nutrient-depleted, and susceptibility to plant disease and pest attack had increased. In Germany, farmers had been able to use seeds from their own cereals (such as rye, wheat, oats, barley) year after year. Now, they found they had to keep introducing new varieties, which might only last a few years. In earlier times it had been possible to grow and cut lucerne on the same field for up to thirty years. At the beginning of the twentieth century, however, that had fallen to just five years.

Out of concern at this degeneration of seed varieties and a decline in food quality, a group of farmers approached Rudolf Steiner for advice. Steiner's response was to give the eight lectures of the Agriculture Course. These were held in June 1924 on the estate of Carl Wilhelm Count von Keyserlingk and his wife, Johanna Countess von Keyserlingk, in Koberwitz, which was then part of the German province of Silesia but is now part of Poland.

Rudolf Steiner and Count Keyserlingk met after the First World War and spoke a great deal about the industrialisation of agriculture, the increasing use of chemical pesticides, and the damaging effects these were having on crops, livestock and people. Steiner was approached to share his insights into farming, and Count Keyserlingk invited him to Koberwitz to give the lectures that laid the foundations for biodynamic agriculture.[1]

Figure 1.1. The Koberwitz estate where the Agriculture Course was held in 1924. It belonged to Carl Count von Keyserlingk and Johanna Countess von Keyserlingk.

Count Keyserlingk's nephew, Alexander Count von Keyserlingk, managed the 7,500-hectare (18,500-acre) estate at Koberwitz. He recalls meeting with Rudolf Steiner in Dornach to set a date for the Agriculture Course:

"I was with [farmer and landowner] Ernst Stegemann on a number of occasions together with Uncle Carl when it was suggested that Rudolf Steiner should be asked to hold a conference for farmers. Stegemann had at that time been working with indications given by Rudolf Steiner, although he had no compost preparations then.

We were only too aware that the future prospects for arable soils, crops and also for human beings and animals as a result of chemical fertilisers were looking increasingly bleak despite the initial high yields that had at first hidden the true results. Due to the increasing nematode population, less and less sugar beet could be grown – and there was no treatment available.

When we saw from Stegemann how well Rudolf Steiner understood the situation, we were keen to learn more. We had no idea then that it would result in a whole course or what wide perspectives Rudolf Steiner would include. We thought he might give some suggestions to counter the destruction of soil structure and the declining quality of our crops. Uncle Carl and Aunt Johanna must surely have already spoken to him about it before I came to Koberwitz.

On the occasion of a visit to Dornach, my uncle Carl gave me the task of extracting from the doctor a time frame for presenting new guidelines for agriculture. On arriving in Dornach I went straight to the Schreinerei and said to Frau Dr Steiner that I would like to speak with the Herr Doctor. I did not have to wait long and when he came out I explained my mission. He responded immediately: 'Yes, I will come to Breslau and give lectures on agriculture.' But I then said: 'Herr Doctor, that is not enough for me. I was not asked to find out *if* you will come but *when* you will come.' Then he chuckled, took out his notebook, turned over a few pages and said: 'Tell your uncle that I will come to you at Whitsun.'"[2]

Preparing for the course

Rudolf Steiner prepared thoroughly for the course, whilst still maintaining his busy schedule. He saw the lectures as building on the centuries-old experience and agricultural practices of local farmers and the science-based agronomy of the time, infusing them with fundamentally new approaches and indications drawn from spiritual science. He spoke with farmers and occupied himself with their concerns. He sought out significant literature to do with arable cropping, fruit production, plant nutrition, soil fertility and agricultural chemistry, and wrote down excerpts and insights in his notebook. He commissioned scientists to undertake trials, and he helped to evaluate and present their findings.

Central to the biodynamic method are the various compost and spray preparations that help improve the vitality of the soil and the health of plants. Among these is the horn manure preparation. Cow horns are filled with manure and then buried in the ground and left over winter. They are dug up in the spring and the manure is then dissolved in water through a vigorous stirring process. The mixture is then sprayed on the soil.

Horn manure preparation, also known as preparation 500, was made for the first time in 1923 at the Research Institute at the Goetheanum in Dornach, Switzerland. Ehrenfried Pfeiffer, one of the scientists commissioned by Steiner who worked at the Institute, provides an account from the time:

In 1923 Rudolf Steiner described for the first time how to make the biodynamic compost preparations, simply giving the recipe without any sort of explanation – just 'do this and then that'. Dr Wachsmuth and I then proceeded to make the first batch of preparation 500 [horn manure]. This was then buried in the garden of the Sonnenhof in Arlesheim, Switzerland. The momentous day came in the early summer of 1924 when this first lot of 500 was dug up again in the presence of Dr Steiner, Dr Wegman, Dr Wachsmuth, a few other coworkers and myself...

Dr Steiner called for a pail of water and proceeded to show us how to apportion the horn's contents to the water, and the correct way of stirring it. As the author's walking stick was the only stirring implement at hand, it was pressed into service. Rudolf Steiner was particularly concerned with demonstrating the energetic stirring, the forming of a funnel and the rapid changing of direction to make a whirlpool ... Brief directions followed as to how the preparation was to be sprayed when the stirring was finished. Dr Steiner then indicated with a motion of his hand over the garden how large an area the available spray would cover. Such was the momentous occasion marking the birth-hour of a worldwide agriculture movement.[3]

A sensitive copper chloride crystallisation procedure was also developed to test quality. Steiner had previously worked with a vet on a treatment for foot and mouth disease, and he gave Lili Kolisko, another anthroposophical researcher, the task of producing and working out the doses in her laboratory at the Goetheanum.

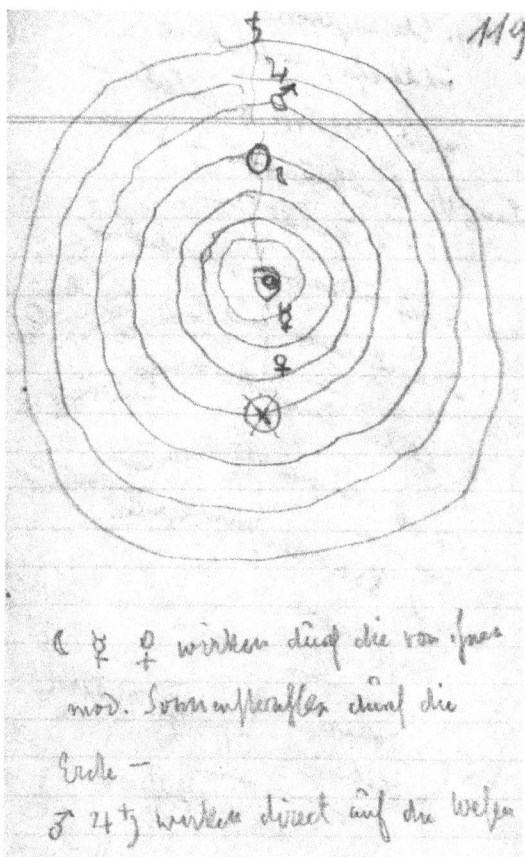

Figure 1.2. A page from Rudolf Steiner's notebook showing a sketch of the solar system with the five classical planets. Writing reads: 'Moon, Mercury, Venus work via the moderated rays of the sun through the earth. Mars, Jupiter, Saturn work directly on life.'

Prior to the Agriculture Course, farmers submitted questions to Rudolf Steiner. They were concerned with farm and plant cultivation, soil, genetics, nutrition, degeneration and disease:

- What are the risks of compromising with new attitudes to agriculture in order to meet the demands of a market in which the focus is on quantity rather than the quality of agricultural products?

- Can the frequent lack of success in fruit production be traced back to the propagation system or the incorrect fertilising and cultivation of the soil?

- Is there an ideal ratio between woodland, arable fields and grassland?

- What nutritional consequences are associated with indigenous and foreign plant products?

- How should the problem of pest control be approached?

- What deeper insights can anthroposophy offer regarding the appearance of plant and animal pests?

- How should cosmic influences be considered when sowing crops?

- How should the farmer behave towards weeds and their control?

- What are the key principles of a crop rotation?

- ✹ How should the farmer best combat plant disease?

- ✹ Is the vitamin cycle linked to the nitrogen cycle?

- ✹ Is the vitamin cycle hampered by mineral fertilisers?

- ✹ What are the specific qualities of the various organic manures (farmyard manure, compost, night soil, liquid manure, waste materials and green manure)?

- ✹ What part do pulses play in human and animal nutrition?

- ✹ How can the symptoms of degeneration in our domestic animals and crops be counteracted in a safe way?

- ✹ What is the reason for the rapid deterioration of potato varieties in particular?

- ✹ What role do soil bacteria play in this cycle?

- ✹ What is the connection between the effect of the smallest entities and the vitamin theory?

- ✹ How do potentised organic substances affect plants?

Preparations for the conference started in the spring. Although 130 people would take part in the course, the number of guests in attendance each day was often a lot more. The staff were given uniforms and were kept incredibly busy with various tasks. The midday meal was mostly eaten standing up since the dining room had to be cleared for the lectures.

Behind the scenes at the Koberwitz estate, the staff were kept very busy catering for the large numbers of people in attendance. Paula Eckardts, a member of the service staff, recalled what it was like:

"There was chaos throughout Germany at the time. There were riots in Sachsen, reds [Communists] were in Hamburg, the Brownshirts [Nazis] in Munich, and we had inflation. Here in Koberwitz it was paradise, an island of peace and order; everywhere else there were rebellions!

Many of those coming to the course from Breslau arrived for breakfast. Once the doors closed for the lecture, we started working like there was no tomorrow. Everything was brought up that was needed to feed this many people. There were sausages and canapés that we had made for the breaks, great mountains of them, and as much to drink as anyone wanted. Dr Steiner would sometimes smile with amazement at the speed with which the many trays were emptied ... In the afternoons everyone left for Breslau, including Dr Steiner, and when they came home in the evenings there was another hot meal. Everything worked out!"

The lectures of the Agriculture Course took place at midday to accommodate those travelling in from Breslau. Each evening everyone went to Breslau, where a Whitsun conference for five hundred people had been arranged. Rudolf Steiner gave lectures on anthroposophical themes in the main hall of Breslau Concert House or in the Assembly Rooms of the Augusta Lyceum, and there was also a public eurythmy performance in the Lobe Theatre and a performance of Goethe's *Iphigenia in Tauris* in the Trade Union Hall, both directed by Marie Steiner. Rudolf Steiner also made himself available for four question-and-answer sessions with farmers, smallholders, farm workers, landowners, tenants and farm estate managers. He had meetings with a youth group for whom he held lectures and led discussions, and on two occasions met with a doctor to offer advice for his consultations.

Reflections of course participant Helmut Woitinas:

"As a young gardening assistant seeking meaning in life and a spiritual home, I was drawn to the *Wandervogel* movement [the German youth movement] ... I then participated in the Agriculture Course. This brought me together with people that I would otherwise have had difficulty getting to know at this time of heightened tension, strikes, unemployment, inflation and a Germany shaken by political radicalisation."

The Experimental Circle

It was during the Agriculture Course that the Experimental Circle of the Anthroposophical Society was founded (it later became the Experimental Circle of Anthroposophical Farmers). The purpose of this group was to coordinate the practical experimental work carried out in the various regions, to produce and distribute the preparations, and to discuss research projects and arrange conferences. It also produced a newsletter. In this way an association of people and farms was created that could take responsibility for developing the indications given by Steiner during the course.

But the founding was not without its challenges. There were disagreements over what approach to take. Some participants favoured a more inner, esoteric approach, while others wanted a greater focus on economics or simply sought practical, day-to-day instruction.

Count Keyserlingk, speaking on behalf of farmers, suggested that the 'uneducated farmers' should simply carry out what the 'enlightened leaders of the [Natural Science] Section' in Dornach decreed. Rudolf Steiner, however, in the address he gave to the course participants that led to the founding of the Experimental Circle, disagreed: 'Thus from the outset we need really active colleagues, not merely people who implement what is proposed.'[4] Rudolf Steiner emphasised his high regard for so-called 'peasant wisdom', which he said penetrated deeply into the cosmic-terrestrial relationships that exist at the place where the farmer is working. Science, by contrast, is in danger of degenerating into dead, abstract knowledge.

The emphasis on this kind of practical knowledge and practical research based on the 'spiritual scientific recommendations for a healthy agriculture' is part of biodynamic agriculture. It is not about theory or hypotheses, nor is it about recipes that merely need applying. Rather, it has to do with much wider perspectives that serve as a foundation for agricultural work. Steiner specifically called for this farmer-based knowledge to be present in Dornach:

> We must, as it were, grow far more together, and in Dornach there must be as much peasant wisdom as can prevail there in spite of its scientific spirit; while science that goes forth from Dornach must be of such a nature that it enlightens the most conservative farmer.[5]

Despite these differences, however, a network was formed and practical work started immediately after the course. From the very beginning there was an ongoing exchange between the Experimental Circle and the Natural Science Section at the Goetheanum. Following the death of Rudolf Steiner, the development of new research was undertaken with the Section through joint conferences and publications, and it was largely due to the Experimental Circle that such a powerful impulse emerged from the Agriculture Course.

Like all other biodynamic initiatives, the Experimental Circle was prohibited by the Nazi regime in 1941. After the Second World War, the work started up anew and, in 1946, a successor organisation was founded: Forschungsring für Biologisch-Dynamische Wirtschaftsweise (the Research Circle for Biodynamic Farming). The Forschungsring was responsible for setting the production standards for Demeter certification, and since 2007 it has devoted itself to biodynamic and organic research.[6]

The question of how scientists and agricultural practitioners should work together remains as current as ever. There is often a tense relationship. New insights are not always put into practice as farmers often prefer to continue as before. And what practitioners know from their own experience is not always recognised by scientists, who consider such knowledge to be purely subjective. What is needed to meet the challenges facing the farming and food sector, however, is precisely the mutual recognition and encouragement of both practice and science that Steiner recommended in 1924.

Figure 1.3. Trial plots set up by members of the Experimental Circle to carry out practical research into Steiner's agricultural methods.

From the Address given by Rudolf Steiner during the Agriculture Course, which led to the founding of the Experimental Circle on June 11, 1924:

"When I look back over my life the farmer who is most valuable is not the big farmer but the small farmer who worked on farms as a peasant boy. If all this is not to happen on a larger scale and is transplanted into scientific terms, it will – as one says in Lower Austria – have grown out of the skulls of peasants. Such an achievement will mean more to me than what I have subsequently undertaken. You should therefore look upon me as this small farmer who has developed a love for agriculture, who recalls his peasant-farming background and can thereby really understand what is living now among this so-called peasantry. You can be assured that this will be understood in Dornach.

I have always been of the opinion – and it is not so ironically meant as it seems to have been understood – that the alleged stupidity or foolishness of farmers is wisdom before God, before the spirit. I have always considered that what peasant farmers thought about things to be much more sensible than what scientists have thought. I have always found it to be so, and it is also the case today. I would prefer to listen to everything that someone working directly with the soil may chance to say about his experiences than to all the ahrimanic statistics which emerge from science; and I have always been glad when I have been able to hear something of this kind, because I have always found it extremely wise. And it is particularly in the area of practicalities, of practical implementation, that I have always found science to be incredibly stupid. And our endeavour in Dornach is to imbue everything undertaken by this science with good sense, which will be achieved by bringing some peasant 'stupidity' into science; and this stupidity will then become wisdom before God.

If we have the will to work together in this way, this will be a really conservative but also thoroughly radical and progressive beginning. It will evermore remain for me a very beautiful memory if this course becomes the starting point for the entry of a genuine, wise peasant quality into the methods of science, which have perhaps not become stupid (that would be insulting) but have certainly become dead."[7]

2. Key Concepts and Practices of the Agriculture Course

We need to acquire new knowledge in order to enter into the whole wider context of things. Humanity has no other choice except either to learn something again in all the different areas of life from the whole of nature, from the whole context of the world, or to let both nature and human life degenerate and die. Just as in times of old it was necessary that people had knowledge that really permeated into the inner workings of nature, so are we also in need of such knowledge again today.

Rudolf Steiner, June 10, 1924[1]

In the hundred years since Rudolf Steiner gave the Agriculture Course, his insights into a sustainable and productive form of agriculture have been adapted and developed in many ways. Indeed, they may be considered a significant impulse behind the growth of the organic movement in the second half of the twentieth century. The biodynamic movement, which arose from it, is now practised in more than fifty countries and on all continents where agriculture is possible.

The farm as a living whole

One of the key ideas of biodynamic agriculture is that of the 'farm organism' as a living whole. Rudolf Steiner even referred to it as an 'individuality'.

The farm organism is understood as the productive working together of soil, plants and animals as they are cultivated by human beings. During the twentieth century, the industrialisation of agriculture splintered this organism into its separate parts, as farming became more specialised. Unfortunately, it is only in the specialised fields of dairy, arable, fruit or vegetable production that a profitable return on the investment of labour, capital and know-how is to some degree possible. The internal exchange of services between the various branches of a mixed farm – such as manure from livestock, straw from arable crops – is no longer viable and specialised farms must buy in such resources from outside. This led to the increasing use of artificial fertilisers, chemical pesticides and herbicides, which completed the industrialisation of agriculture. This occurred on a grand scale in the twentieth century, bringing with it both progress and problems.

The biodynamic movement has always seen itself as a place where the living agricultural whole is cultivated in all kinds of ways, despite real economic challenges. Each generation must learn to appreciate the value of this.

The farm organism

The model for thinking of the agricultural entity – whether it is a holding, a village or a whole valley – as an organism can be found particularly in mammals. Here each single bodily organ exists to serve the whole and derives its main purpose from it: we might think of the heart, lungs, liver and kidneys in human beings. Similarly, the various parts of the farm are seen as organs of the larger farm organism.

For example, the herd of cows is no longer simply seen as a source of income gained through the sale of milk. Instead, through the effect of grazing on the landscape and through using the manure they produce, the herd of cows becomes the metabolic pole of the whole organism. If these effects and resources are kept in mind, then the approach to keeping and raising animals will be different.

The focus of the farmer (or the farm community) is primarily on the farm as a whole – that is their main entrepreneurial task and function. The formation and management of the individual elements are secondary, and these must always accord with the requirements of the overarching organism. This orientation around the whole must not be pursued in an absolutist way, however. The farm must after all be able to survive and respond immediately to changing market conditions. The challenge is to meet these demands without being entirely dependent on them. The farmer's keen observation and management of the farm allows room for manoeuvre. It also helps to prevent a one-sided dogmatism from taking hold, which Steiner warned against.

The core principle of the farm organism is that it be self-contained. This is made possible by wide internal diversity and a closed cycle of substances: manure from the animals, which is spread on the soil to grow the fodder to feed them, and which then produces more manure. In the Agriculture Course, the aim of achieving as closed a cycle as possible is presented as the pre-condition for growing produce of the right quality. Biodynamic agriculture, like any other form of organic agriculture, is not concerned with the endless recycling of dead materials, but with substances that are thoroughly permeated with the life and nature of the place where they belong. Take nitrogen, for example, which has a completely different quality when released from the internal cycle of substances than it has when applied externally as an artificial fertiliser in a highly soluble form.

The manuring or fertilisation question is, for Steiner, one of 'the most important questions in agriculture', and it is on this point that he expands the idea of the farm organism as a living whole to include the concept of the 'farm individuality'.[2]

The farm individuality

With the idea of the 'farm individuality', Rudolf Steiner introduced a concept into farming that broke with the conventions of classical agronomy. With this idea, the human being as an individual with a unique biography is taken as a model for the farm. Every farm, every farm community, has its own many-layered character, a specific quality, and its own history. It is influenced by its situation, by its environment and the cycle of the year. In this way, biodynamic agriculture takes on the colour and diversity of individual development and practice. It is no standardised scheme with fixed recipes and standards set in stone, rather it is an invitation to work creatively, and a way of engaging responsibly with the earth entrusted to our care.

From Rudolf Steiner's address to the Experimental Circle, June 11, 1924:

"I have spoken about the fact that an estate or farm is always an individuality (in the sense that it is never the same as any other). Climate and soil conditions provide the essential foundation for the individuality of the farm. A farm in Silesia is not like the one in Thuringia or southern Germany. They really are individualities. Now it is the prevailing view of anthroposophy that generalities and abstractions are of no value whatsoever and they are of the least value if one wants to engage in practical realities."[3]

Fodder, dung and seed chaos

The flow of substances circulating through the farm arrives as fodder for the animals, in particular the cows. While chewing the cud they carry out what Steiner calls a 'cosmically qualitative analysis'[4] This means that as it is eaten and digested, the feed is 'perceived' physiologically by the ruminating cows. The result of this 'analysis' – that is, what the animal experiences in the unique qualities of the fodder, the particular soil and climatic conditions of the place where it was grown – is then imprinted on the substance.

During the digestive process the fodder is converted into dung. The manure that comes from the cow is of the same substance as the fodder, only now it is thoroughly permeated by the forces impregnated into it by the cow. If this manure is used as fertiliser, not only will it provide nutrients for the plants but also the formative forces particular to that location and therefore most suited to it. This is then used to produce the following year's fodder or is organically integrated within the whole farm via crop rotation. This leads over the years to an ongoing 're-encountering' process: the unique qualities of a place are enhanced, which brings about a process of individualisation.

Creating a closed cycle of substances is what opens up the farm as an organism to the creative potential of a developing individuality. Whenever one cycle of life comes to an end and a new one begins, an opening occurs for the creative forces of the cosmos to pour in. This is referred to in the Agriculture Course where Steiner describes how the seed is brought into a condition of chaos that allows the surrounding universe to work in and stamp itself on the forces of the seed. This leads from one plant generation to the next:

> The earthly process of organisation is brought every time again to an end even to the point of chaos when a seed is formed. The new organism is built up every time again out of the whole universe in the chaos of the seed.[5]

This means that over the regular course of time, the closed cycle of substances that human beings have created by carefully processing farmyard manure periodically opens itself up to the eternal and the timeless: to the creative, ever-new forces pouring in from the universe. From this it becomes apparent that the closed organism is not an end in itself. Rather it lays the foundation for an *inner* opening, an awakening of the individuality of the living farm as a whole. The biodynamic spray preparations also serve to widen and incorporate the quality of time in agriculture in a more dynamic way.

The soil between the worlds above and below

In the second lecture of the Agriculture Course, Rudolf Steiner compares the farm to an inverted human being. The upper part, the head in the case of the farm individuality, is below in the earth, and the lower part, the metabolic-limb system of the farm individuality, is above in the light and air. In the middle, where the upper and lower realms meet, lies the soil. This can be compared to the human diaphragm, an organ that breathes with the different rhythms where the upper and lower realms meet.

The farm therefore not only exists on a horizontal level, measured by the number of hectares or acres. For biodynamic practitioners it also has a vertical dimension. And if we ask after the upper and lower boundaries of the farm, no measurable limitations can be given. With its geotropic growth the root of the plant is oriented towards the centre of the earth and its shoot and flower to the sun. As a result, the terrestrial and cosmic processes of life interpenetrate one another on the farm. This interpenetration can be far more intense in the cultivated organism of the farm, where the closed cycle of substances allows the relevant forces to work more strongly. The compost preparations, which bring together elements and forces from both the horizontal and vertical dimensions of the farm organism, are the new and direct means of putting this aspect to practical use.

The earth's surface as a diaphragm:
"If a study of the soil is taken as a point of departure, the first thing to notice is that the soil is a kind of organ within the organism which manifests itself in nature wherever there is growth. The soil is a real organ, it is an organ that we might well compare with the human diaphragm ... When on a farm we are actually walking around in its belly, and the plants grow up into this belly of the farm."

Lecture Two, June 10, 1924[6]

The strength of individual identity

A place that has been developed as a living whole and cared for over many years – a farm, a garden, a park or a valley – creates within itself all the elements found in nature. It is in this relationship between the specific and the universal that the identity of a farm is found. The farm's strength of identity, which grows ever more important for the success of the farm, is an expression of the farm individuality. What does this have to do with the people working on the farm? This question is frequently discussed in the biodynamic movement and has led to a wide range of views depending on each individual's standpoint. Engaging with this 'I–You' relationship between human beings and the farm individuality is one of the most essential and exciting challenges of biodynamic agriculture. From another angle it is interesting to discover how the individuality of a small farm relates to planet earth as a whole. Just as the individual is the place where the future of humanity will be realised, so the farm, which is developing towards individuality, is the place where the future of the earth will be realised.

The five sisters

In the third lecture of the Agriculture Course, Rudolf Steiner discussed the substances that form protein as a carrier for life on earth and allow what is spiritual to work down into what is material. He referred to these substances as 'sisters'.

- **Sulphur** acts as the mediator between the formative power of the spirit in the universe and what is physical. It is the 'bearer of the spirit', according to Steiner.[7]

- **Carbon** builds the framework of the organic world and lives dynamically in the life cycle of each plant. If carbon can be retained within the cycles of life, it will not only cause no harm but will have a positive effect on the climate.

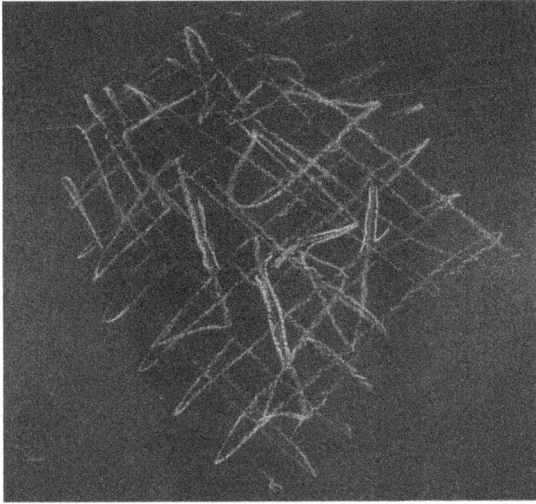

Figure 2.1. Representation of 'carbon-like scaffolding' that underlies all living things. Blackboard drawing by Rudolf Steiner, Koberwitz, June 11, 1924.

⚘ **Oxygen** is the bearer of life and works together with sulphur to carry the influence of life out of the spiritual universe into the physical.

⚘ **Nitrogen** is the mediator between oxygen and carbon, or to put it another way, between *life* and *form*. Wherever the principle of life seeks embodiment in form, there you will find nitrogen working.

⚘ **Hydrogen** is the bearer of what was alive and formed into the universe. Hydrogen dissolves everything away and 'carries everything of a formal, enlivened astral nature back into the expanses of the universe so that it becomes possible for it to be taken up again by the universe' – that is, through the sulphur process that carries what is spiritual into the physical.[8]

Nitrogen from the air is hard to bring into the organic realm. Animals and members of the legume family can, however, make sufficient nitrogen of the right quality and quantity available to the farm. Biodynamic agriculture and the Demeter standards require livestock to be included to further facilitate this. Synthetic and industrially produced nitrogen fertiliser is unnecessary. This artificial nitrogen quickly evaporates into the atmosphere, where it is 265 times more damaging than CO_2.

Nitrogen and... nitrogen:
"No one knows today for example that all types of mineral fertiliser are precisely what is causing this degeneration that I have spoken about, that contribute significantly to the decline in quality of agricultural products. Everyone simply believes today that a certain amount of nitrogen is needed for plants to grow, and people don't think it matters how this nitrogen is prepared or where it comes from. It is not a matter of indifference where it comes from, however. The fact is that between nitrogen and nitrogen, between nitrogen as it appears in the air together with oxygen, between this dead nitrogen and the other nitrogen there is a huge difference. You will not deny, my dear friends, that there is a difference between a person who is alive and moves around and a corpse, a human corpse."

Comments made by Rudolf Steiner on June 20, 1924,
following the Agriculture Course

The preparations

Central to biodynamics are the preparations that are used to improve the vitality of the plants and the soil, and which Steiner introduced in the fourth and fifth lectures of the Agriculture Course.

Easily soluble mineral fertiliser salts cannot promote life, yet traditional farm fertilisers and purely natural organic processes are not always able to replace the forces taken from the soil because of the harvest. This insufficiency can be compensated for by dynamising the soil with compost that has been treated with the preparations. The two spray preparations, horn manure and horn silica, encourage 'healthy root growth and enhance ripening',[9] and the six compost preparations help to 'stabilise the soil's nitrogen content, promote microbial diversity, and regulate the decomposition and humus-forming process'.[10]

The preparations are sometimes referred to by a designated number, beginning with horn manure, known as preparation 500, and horn silica, known as preparation 501. The compost preparations then follow in sequence from 502 to 507.

Horn manure (500) and horn silica (501)

The two spray preparations, horn manure and horn silica, are applied to the soil and the plants respectively. Manure is used to fill cow horns and is then buried in the ground over winter. What is formless is exposed to the crystallising forces of the solid and liquid elements. The horns are dug up in the spring.

Similarly, in a complementary fashion, silica or rock quartz is ground to dust and water is added to create a kind of paste that is also used to fill cow horns. These are then buried in the ground over summer. What once possessed a definite form is now formless and exposed to the airy warmth that pervades the soil during this time. The horns are dug up in the autumn. Both substances are then diluted separately in water through a vigorous stirring process.

The horn manure solution is sprayed on the soil in the evening, just before sowing, and is used to promote healthy soil and root growth. The horn silica solution is sprayed on the plants in the morning and is used to improve the quality of leaf, flower and fruit formation.

Yarrow (502)

This preparation works with the sulphur and potassium processes that form protein and help to build up the structure of the plant. Yarrow flowers (*Achilea millefolium*) are collected, moistened and then placed in the bladder of a stag. The bladder is hung in a place that is exposed to the sun during the summer and is then buried in the ground in autumn. It is dug up in spring and the humus-like substance is used, in homeopathic amounts, to inoculate the manure or compost with it.

Chamomile (503)

The second preparation works with calcium compounds, binding them together with 'other substances necessary for healthy plant growth'.[11] It also helps to stabilise the nitrogen content in the compost. Chamomile flowers (*Matricaria chamomilla*) are collected and used to fill the small intestine of a slaughtered cow. These chamomile sausages are also buried during the winter and dug up in the spring. A small amount of the preparation is then placed in the compost heap along with the other preparations.

Stinging nettle (504)

According to Steiner, using this preparation is like infusing the soil with intelligence.[12] It works with potassium, calcium and iron and organises these processes in a more effective way to bring health and harmony to plant life. The stinging nettles (*Urtica dioica*) are cut before flowering and allowed to wilt. They are then buried in the ground without an animal sheath and left there for a whole year, before being dug up and applied to the compost heap.

These first three preparations can be seen as working in the earth in the 'head' of the farm organism, drawing closer to the surface of the soil, which, as Steiner describes it, forms the diaphragm of the farm individuality.

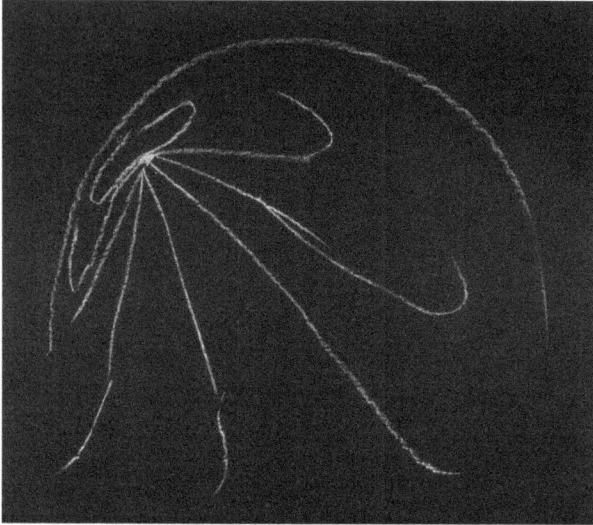

Figure 2.2. Representation of the radiating effects of the compost preparations in a compost heap. Blackboard drawing by Rudolf Steiner, Koberwitz, June 13, 1924.

Oak bark (505)

This preparation supports the health of the plants, working with calcium to fight plant disease. Fresh oak bark (*Querqus robur*) is finely grated and stuffed into the skull of a domestic animal. The skull is then buried over the winter period before being dug up and added to the compost heap.

Dandelion (506)

This preparation works with silicic acid to bring it into the right relationship with potassium. Dandelion flowers (*Taraxacum officinale*) are collected and dried in spring. They are then used to fill bovine mesentery, the membrane that surrounds the abdominal and pelvic cavities and all the organs they contain. This is then buried over winter before being dug up and added to the compost heap.

Valerian (507)

This preparation supports phosphorus processes. Valerian flowers (*Valeriana officinalis*) are collected and squeezed, and the juice is stored in bottles in the sun. Like the other preparations, it is added in minute quantities to the compost pile. It covers the pile in a kind of protective sheath, rather like a skin that turns the pile into a self-contained entity.

The interconnectedness of nature:
"People are accustomed today to look at a plant as something in its own right, and then at one plant species and another beside it as objects quite apart from one another. They arrange everything beautifully in pigeonholes, in separate species and genera, according to what is supposed to be known of them. But this is not how it is in nature. In the world of nature everything is in a state of mutual interaction; one thing is always influenced by another. In our present materialistic age we focus our attention only upon the coarser kinds of interaction of one thing upon another, thus when one animal is eaten and digested by another or when dung from animals is spread on the fields. Attention is

devoted only to these more obvious kinds of interaction. However, there are in addition interactions that take place as a result of more refined forces."

Lecture Seven, June 15, 1924[13]

3. The Historical Impact of the Agriculture Course

This chapter looks at the impact the biodynamic movement has had on the wider organic movement, how it has influenced developments in social and organisational structures for farming and marketing biodynamic produce, and at research carried out into biodynamic methods and how they compare with organic and conventional agriculture. Finally the chapter looks at biodynamics in relation to the traditions of Indigenous people and how it can be seen as a cultural impulse.

The biodynamic movement and the rise of organic agriculture

The term 'biodynamic' did not originate with Steiner. Instead, reference was initially made to 'biological fertilisation'. Ernst Stegemann, the landowner who had worked with Rudolf Steiner's early indications for a new agricultural method prior to the Agriculture Course, favoured the term 'dynamic farming'. The anthroposophist Erhard Bartsch, who later founded the Reich League for Biodynamic Agriculture, preferred 'organic farming'. Through a synthesis of these two names, the term 'biodynamic' was created.

Although biodynamic agriculture developed within the anthroposophical movement, it is also part of the wider organic movement, which developed during the 1930s in response to the increasing mechanisation of mainstream agriculture and the use of chemical fertilisers. It was inspired by the work of a number of people, among them Sir Albert Howard, Lady Eve Balfour and Lord Northbourne. But Rudolf Steiner and his co-workers were far and away the first to chart a new path for agriculture.

Sir Albert Howard: composting pioneer

One of the early pioneers of the organic movement was the English botanist Sir Albert Howard. Howard was an agricultural advisor to states in central India and the director of the Institute of Plant Industry in Indore from 1924 to 1931. He had been sent to India to teach Western agricultural methods but soon discovered that he could learn more from the Indians themselves. In particular, he noticed the connection between healthy soil and the health of the local population, livestock and crops. Howard became a supporter of Indian sustainable farming and was the first Westerner to document their techniques, which he published in his book *The Waste Products of Agriculture* in 1931. The book was concerned with the destruction of the earth's soil and humus reserves, which led Howard to develop the Indore composting method. Based on Indian composting techniques, this method alternates layers of nitrogen-rich 'green' materials with carbon-rich 'brown' materials to produce a high-quality compost that is rich in humus.[1]

Howard's encounter with biodynamics came when he met Ehrenfried Pfeiffer at a science conference in Britain. He learned more about biodynamic methods during the 1930s from their early starts in both Holland and Britain, but he remained unconvinced. In the foreword to his 1940 book, *An Agricultural Testament*, he wrote:

> I [am] sceptical about whether the followers of Rudolf Steiner can give a true explanation in accordance with the laws of nature or can prove their theories with any practical examples.[2]

With the outbreak of the Second World War, this tentative connection between the organic and biodynamic movements was put on hold.

Howard's achievements and those of his first and second wives, Gabrielle and Louise, were known to only a few specialists in Europe at the time. Even the biodynamic movement had to rediscover his pioneering work, spreading out across India from those places where he had conducted his research and trials. In India, however, this origin of organic agriculture has never been forgotten.

Marienhöhe: a model biodynamic farm

Georg Michaelis, lawyer, Prussian official and, briefly, Chancellor of the Weimer Republic, was initially sceptical of anthroposophy. However, he changed his mind after his son-in-law, Martin Schmidt, attended the Agriculture Course in Koberwitz. Michaelis went on to become the co-founder of

Demeter, the organisation that certifies biodynamic products and produce according to strict standards.

In 1927, Michaelis helped a group of anthroposophists, led by the farmer Erhard Bartsch, to purchase Marienhöhe farm east of Berlin. It soon became a model farm. During the 1930s it was the centre for biodynamic agriculture in Germany and the development of the Demeter standard. Demeter is used to this day as a trademark for products certified as following biodynamic production and processing standards. Mandatory principles have since been established for marketing and for social standards throughout the supply chain.

Michaelis was able to draw on his experience as an official in the ministry of finance and the imperial institute for grain and national food reserves to offer advice and instruction. He also helped to establish the Verwertungsgenossenschaft Demeter (Demeter Food Processing Cooperative) and later the Demeter-Wirtschaftsbundes (Demeter Economic Association).

Figure 3.1a. Marienhöhe farm before it was purchased by Georg Michaelis in 1927.

Figure 3.1b. Marienhöhe farm in 2017.

When biodynamic work was threatened by the Nazis, Michaelis, who was already a member of the party, entered into negotiations with Rudolf Hess, Hitler's deputy, and Walter Darré, director of German agriculture policies. Darré was sympathetic towards organic and biodynamic agriculture and was opposed to the use of artificial fertilisers and pesticides. However, the Nationalist-Socialist affiliated farmers' union in Bavaria, which supported the use of mineral fertilisers, protested against the biodynamic approach. They declared it and its manuring methods to be an evil product of the internationally oriented views of anthroposophy and its supposedly Jewish founder.

Biodynamic agriculture in the Third Reich

In November 1935 the Anthroposophical Society was banned in Germany and the few thousand members of the society were registered and put under surveillance. Some members were detained and interrogated. Although the anthroposophical movement was considered an enemy of the Nazi regime and antagonistic towards its ideology, most of the anthroposophical institutions, such as Waldorf schools, children's homes, medical practices and farms, were allowed to continue working so long as they didn't engage in political opposition or openly promote anthroposophy.

To begin with these institutions benefited from the tolerance of high-level party members. A few anthroposophists with a special grasp of biodynamic agriculture were employed on farms run by the SS. In the end, however, the Reichsverband für Biologisch-Dynamische Wirtschaftsweise (National Biodynamic Association) was prohibited in June 1941 as part of a campaign against secret teachings and occultism. Anthroposophical literature was also seized and some members of the association were imprisoned. Heinrich Himmler, who, like the director of German agricultural policy, Walter Darré, was wary of the increasing industrialisation of agriculture, ordered fertilisation trials to be undertaken, which included a biodynamic variant. The trials took place under the auspices of the SS on the farming estates of the Deutschen Versuchsanstalt für Ernährung und Verpflegung (the German Research Centre for Food and Catering), which the SS had founded in 1939. The plan was to see if biodynamic methods could be used to serve the aims of the regime, but in so doing they separated the practical measures from the spiritual-anthroposophical view of the world and human beings.[3]

The organic movement in Switzerland

In the 1940s, Dr Hans Müller and his wife, Maria Müller-Bigler, developed the organic-biological movement in Switzerland. They were both familiar with biodynamics and with the idea of closed cycles within sustainable farming, but they had no affinity for the more dynamic aspects relating to the preparations or with anthroposophy itself. As a result, they focused their attention on the biological aspects. They were helped in this by the work of Hans Peter Rusch, whose own research into the microbiological conditions of different types of soil allowed them to develop a scientific approach to tackling the problems of organic farming and maintaining the long-term fertility of the soil.[4] For this group of pioneers, reverence for life as an inner attitude was an important motivation and a reason to avoid chemicals.

After periods of tension between the biodynamic and organic-biological groups, they finally found a way of working together. In 1973, the Research Institute of Organic Agriculture (Forschungsinstitut für biologischen Landbau or FiBL) was founded with the aim of encouraging the growth of organic farming and providing practical advice to farmers who wanted to avoid using conventional agricultural methods. Then, in 1981, Bio Suisse was created as the main organisation of organic agriculture in Switzerland. As an umbrella organisation it included over thirty organic farmers' associations in its membership, including the FiBL.

This is how the insights of Rudolf Steiner and those representing biodynamics contributed to the modern organic movement. Key questions that had been addressed by the biodynamic movement – for example, those relating to plant breeding, animal husbandry, training and socioeconomic issues – became important for the wider organic movement.

In 1978, senior lecturer in environmental psychology Bill Mollison and graduate student David Holmgren from Tasmania published *Permaculture One*. The book presented their research into the unsustainable methods of modern industrialised agriculture and the negative impact it was having on biodiversity. It also introduced permaculture to the world.[5] This essentially green social movement is based on concern for the earth, humanity and the future, and takes a broad approach to improving practical ways of working in fields and gardens.

In 2015, French filmmakers Cyril Dion and Mélanie Laurent made *Demain* (*Tomorrow*), a documentary that identifies various initiatives that solve environmental and social challenges. In the section on agriculture, they included the permaculture movement.[6]

Since 2020 the permaculture movement has been directly sponsored by the Swiss Ministry of Agriculture.

The impact of artificial fertilisers

By the end of the First World War farmers were already applying artificial fertilisers to their fields, but it wasn't until the 1960s that a huge change in farming practices took place around the world.

In Europe, Sicco Mansholt was the Agriculture Commissioner for the European Economic Community (EEC) and was responsible for directing the industrialisation of European agriculture through a subsidy policy. Small family farms were being replaced by large industrial farming units that specialised in livestock, arable crops or horticulture. Within ten years almost all mixed farms had disappeared. It was a dramatic change that occurred not only in Europe, but also in many places across the world, especially in Asia, where the introduction of technology and chemicals in agriculture brought about a higher standard of living. The cultivation of new varieties together with artificial fertilisers and synthetic pesticides resulted in yields almost three times higher than before. But the shadow-side of this agricultural transformation soon showed itself.

When the Club of Rome, an informal organisation of intellectuals and business leaders, published its report *The Limits to Growth* in 1971, it became clear to Mansholt that the policies he had been pushing would have catastrophic consequences for the environment. He radically changed his point of view and for the rest of his life was an avid supporter of organic agriculture.

Half a century later Mansholt was proved right. It has become clear, not only in the realm of policy but at all levels of society, that an agricultural approach that uses large quantities of chemicals and is focused primarily on increasing productivity is extremely harmful to the environment. Intensive livestock farming is a major contributor to greenhouse gas emissions, as is the manufacture and use of nitrogen fertiliser. The decline of biodiversity is primarily due to pesticide use, and the pollution of waterways and desertification is also due to modern methods of agriculture. The growing power of a small number of huge corporations that produce genetically modified (GM) and hybrid seeds, artificial fertilisers and pesticides determines what comes into our shops and onto our dinner plates. Farmers are then caught between the suppliers of seeds and pesticides and the low prices demanded by multinational food retailers. Conventional agriculture is in crisis and coming under fire from all sides.

Wake-up call (I): Rachel Carson and *Silent Spring*

After the Second World War there was growing concern about the harmful impact of toxins on the environment. Biologists and environmentalists were particularly alarmed by the negative effects the supposedly harmless pesticide DDT was having on fish and bird populations among other things. These concerns surrounding the unlimited use of pesticides was brought to public attention in 1962 when biologist, author and environmental activist Rachel Carson published her book *Silent Spring*. This was a seminal moment that marked the beginning of the modern environmental movement and ultimately led to the banning of DDT.

Figure 3.2. Rachel Carson, author of Silent Spring.

During the writing of *Silent Spring*, Carson became good friends with Marjorie Spock, a Waldorf teacher, eurythmist and biodynamic gardener who lived on Long Island, New York. In the summer of 1957, the Department of Agriculture began a massive aerial spraying campaign using DDT with the aim of eradicating the gypsy moth (*Lymantria dispar dispar*). The toxin was mixed with heating oil and applied fourteen times a day. These measures affected 3 million hectares (7.4 million acres) in the north-eastern United States, including the biodynamic land managed by Marjorie Spock.

Marjorie Spock joined the inhabitants of Long Island in their protest against the government along with her friend Mary Richards, anthroposophist, environmental activist and biodynamic gardener. Richards' vegetable garden, the soil of her entire holding and the fauna and flora were being poisoned by this aerial spraying.

Figure 3.3. Aerial spraying for pest control.

Supported financially by their neighbour, the world-famous ornithologist Robert Cushman-Murphy, Marjorie Spock and Mary Richards mounted a legal campaign against the US government to prohibit aerial spraying by the Department of Agriculture. During the three-year-long ordeal, which ultimately met with failure, Marjorie Spock wrote daily progress reports and sent them using a primitive fax machine to various interested and influential friends. Among them was Rachel Carson, who wrote about the case in *Silent Spring*:

The suit brought by the Long Island citizens at least served to focus public attention on the growing trend to mass application of insecticides, and on the power and the inclination of the control agencies to disregard supposedly inviolate property rights of private citizens. The contamination of milk and of farm produce in the course of the gypsy moth spraying came as an unpleasant surprise to many people.[7]

Several chapters from Carson's book were published in the *New Yorker* magazine in June 1962, prior to the book's publication in September. The three-part article angered the authorities and pharmaceutical companies, who branded Carson a communist and accused her of trying to sabotage the food industry. Nevertheless, Carson's book remained on the *New York Times'* bestseller list throughout the autumn and winter of 1962, and by the spring of the following year it had sold nearly half a million copies.

Rachel Carson had published several bestselling books about life in the ocean, but *Silent Spring* was her most successful book. Unfortunately, she did not live to see the impact it had on the public. In April 1964, she succumbed to her long fight against cancer, aged just 57. Marjorie Spock, however, was able to see the effect that Carson's wake-up call had on people: the growing awareness for the need for organic production, and the public's changing attitude towards food. Marjorie Spock died in 2008, in Maine, at the great age of 103.

Wake-up call (II): Philippe Matile and the living soil

On October 22, 1966, an article appeared in the Swiss evening newspaper *Die Tat* called 'Attack on the Foundations of Life: The Limitations of Artificial Fertiliser Use in Agriculture' by Dr Philippe Matile, professor of plant physiology at the Eidgenössische Technische Hochschule Zürich (the Swiss Federal Insitute of Technology in Zurich or ETH Zurich).

In a very matter-of-fact way, the author countered the promotional language of the fertiliser industry:

A massive increase in yields seemed at first to completely justify this fertiliser revolution. The idea that mineral salts removed from the soil by the harvested crop had to be replaced, had – and still does have – something compelling about it. The usual fertiliser advice, based on the analysis of soluble salts in the soil, demonstrates that agriculture right up to the present day remains wedded to the idea that the soil is lifeless …

Since the discovery of purely mineral-based plant nutrition, which led to the idea of soil being dead, independent research has shown that the soil is something that is alive … A science that recognises the reality of the living soil must come to see it as a vast living community of the plant mantle and organism of the earth …

A form of agriculture that works with the living soil knows, therefore, about the great significance of organic fertilisers like manure, compost and sewage sludge etc. … So far, too little notice has been taken of the results of trials carried out over decades on experimental farms using organic and inorganic manures. The truly impressive and scientifically validated results of New Belis farm in Suffolk, for example, have shown that a soil treated correctly under a modern intensively managed system can survive with no artificial fertiliser inputs for 25 years without any reduction in fertility and with the best agricultural outcomes on all levels. The soil is clearly not impoverished if the farmer is able to maintain the optimal activity of the invisible life in the soil through the carefully crafted use of organic fertilisers.[8]

During his postdoctoral period at the ETH, Philippe Matile had lived with his family on Breitlenhof, a biodynamic farm in Hombrechtikon near Zürich managed by Emil and Alice Meier. As a plant physiologist he was amazed to see how well this farm functioned without any additional mineral fertilisers and he drew on these observations in his article.

The article went on to be re-published in several newspapers. On February 1, 1973, Matile co-founded the Swiss Foundation for the Promotion of Organic Agriculture, which laid the foundations for the FiBL.

Ehrenfried Pfeiffer: biodynamic pioneer in America

Ehrenfried Pfeiffer was the young scientist who was present when the first horn manure preparation was developed in 1923. In 1938 his book *Soil Fertility: Renewal and Preservation* was published. At the time, numbered copies of Rudolf Steiner's Agriculture Course were restricted to selected professionals and this continued to be the case until the 1960s. Pfeiffer's book was therefore the first to make biodynamic agriculture available to the wider public in a readable, comprehensive way, although Steiner's name and anthroposophy were hardly mentioned in it.

Figure 3.4. Ehrenfried Pfeiffer (left)

As a result of this publication Pfeiffer was invited to attend the Betteshanger Summer School and Conference on Biodynamic Farming in Kent in July 1939. It was the first conference of its kind to be held in Britain and was organised by Lord Northbourne. Pfeiffer was the lead presenter and over a nine-day period he gave a course on biodynamic agriculture to British farmers. Following the conference, Lord Northbourne wrote a book called *Look to the Land*, which expanded on the ideas presented there. The book was published in 1940 and introduced the term 'organic' into the ecological movement. Unfortunately, due to the outbreak of the Second World War, exchanges between the organic, ecological and biodynamic groups stalled.

Only speak out of experience:
"In the spiritual and in the physical realm we should only speak out of experience. In our close research community, we must look above all towards practical experimentation and avoid all speculative theories. Our people simply have too little laboratory experience and are unable to convert ideas into research trials."

Ehrenfried Pfeiffer in a letter to Alla Selawry,
November 13, 1958[9]

In 1940, Pfeiffer emigrated to America with his family, where he became the leading personality in the development of biodynamics. At a time when little consideration was given to waste and how it might be used, either in Europe or America, Pfeiffer developed an industrial-scale composting system to process city waste and developed his Compost Starter.

Over the coming decades, Pfeiffer continued to play a significant role in the spread of biodynamics through his work as a researcher, advisor and farm manager. As a trained chemist he occupied himself extensively with the chemical and biological issues relating to agriculture, nutrition and medicine. His interest in soil erosion and soil retention, landscape and its interaction with agriculture, and the quality of food and the influence on it of production and processing was focused on what was practical and most useful. In 1957 he was invited to take up a professorship at Farleigh-Dickinson

University in Rutherford, New Jersey. It was the first professorship in 'integrated organic science'.

He ran a laboratory for agricultural and medical research in Threefold farm in Spring Valley and worked as a much sought-after advisor until shortly before his death in 1961.

Maria Thun's Biodynamic Calendar

In his lectures, Rudolf Steiner had pointed out the connection that exists between cosmic forces and plant growth. These indications were later taken up by Maria Thun as she developed her Biodynamic Sowing and Planting Calendar.

Maria Thun had grown up on a small farm near Marburg in Germany and had helped her father with farm work. Thun's interest in biodynamics was kindled when she met her future husband, Walter Thun, and he introduced her to a number of biodynamic farmers. After that, Thun attended various introductory courses at the Institute for Biodynamic Research in Darmstadt. Inspired by Franz Rulni's planting calendar, Thun began experimenting with sowing radishes and noticed that there is a relationship between the position of the moon in the zodiac and the growth of the radishes at the time of planting.

Maria Thun continued her research over the next six decades, eventually establishing the Centre for Constellation Research in Crop Cultivation near Marburg in 1971. She noted the influence of the planetary constellations on the cultivation of soil in fields and gardens, and was able to connect the four classical elements and the constellations with the different parts of the plant. Root vegetables (carrots, beetroot, radishes) were influenced by the earth element, fruiting vegetables (cucumbers, tomatoes, courgettes, peppers, beans) by the warmth element, leaf vegetables (spinach, chard, cabbage, lettuce) by water, and flowers by light and air. Using this information, the calendar indicates when each category of vegetable is best sown, cultivated and harvested.

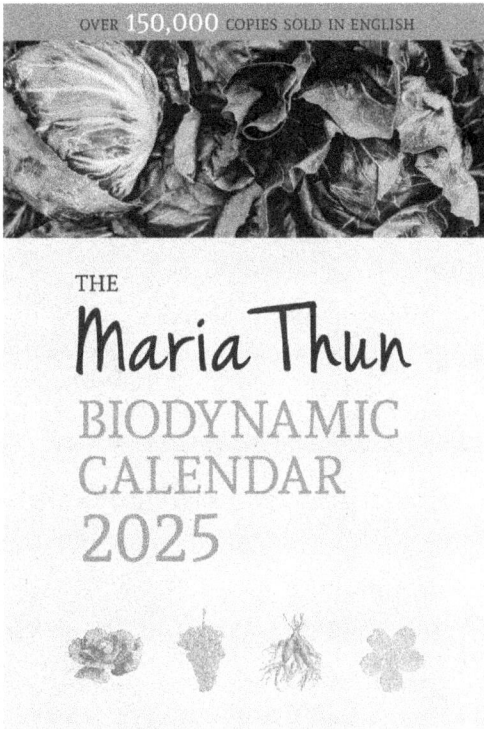

Figure 3.5. The Maria Thun Biodynamic Calendar.

Maria Thun's Biodynamic Sowing and Planting Calendar was first published in 1963 and has been in print every year since then. Drawing on the relevant constellations and moon phases, the calendar gives recommendations and helpful suggestions to gardeners, horticulturalists, farmers, wine growers and beekeepers concerning planting, sowing and harvesting, as well as information on favourable and unfavourable times for working in the garden. The calendar is now available in twenty-seven languages and reaches far beyond biodynamic circles in its popularity and appeal.

The farm as a social laboratory

The concerns of Rudolf Steiner's Agriculture Course extend far beyond the primary issues of agriculture and food production to embrace an enhanced awareness of our relationship to nature and each other, and a different attitude towards consumerism. During the 1970s and 1980s many biodynamic farms became places where new social and economic structures were tried out. Young people from cities developed many forms and structures that could break the chains of traditional land ownership and lead it over into a form of common stewardship.

Questions surrounding the land are of vital importance to agriculture. To whom does the land belong? Who has the right to farm it? How is the relationship negotiated between ownership and the right to farm the land? How is farm succession organised? How is the capital found to purchase land at market prices and then make it available for farming under terms that are affordable? What is the legal and financial entitlement of this capital?

The long-term, intergenerational, legal and financial responsibility for the land and buildings can be transferred to a 'juristic person' (a legal entity that has rights and duties similar to an individual) committed to biodynamics. Ownership can be managed by those running the farm as well as by members who represent the non-farming community of supporters. Released from the stream of inheritance, farmers can develop new collegial ways of working together that do not have to rely on close family members. As a result of this, a whole series of communities and farms have arisen that operate in direct association with other economic ventures.

Organisational structures

The insight and practical experience of a farm conceived of as a single whole – as an organism and an individuality – if it is to be applied effectively, challenges each generation to find the appropriate living structures. It requires an innovative and creative capacity for developing the farm as an organism and for finding the social forms needed to work together. This also includes marketing and ownership of the agricultural capital with its soil, plants and animals.

The classical mixed farm

This type of farm is usually organised around a family that has been living on it for two or three generations and involves the full range of farming activities: soil cultivation, manure and compost management, the production of a wide range of crops and often taking care of many types of livestock. The size is not relevant. It can be a huge farming estate that employs a whole team of farm workers, a typical West European family farm or, as in southern India, a smallholding of just a couple of hectares. Many if not most biodynamic farms operate on this basis. Such farms clearly have certain social limitations, demanding as they do total commitment every day from one year to the next. Life and work are one. In earlier times this was often seen as an ideal, but today the long-term identification of life with work holds much less appeal.

The community farm

From the 1970s onwards, the biodynamic movement saw the founding of farm communities, especially in Germany. Between three and five families would share the different farm activities of tending livestock and arable crops, vegetable production, processing and marketing. As a result, larger farm operations could be managed, and increased specialisation allowed for more tailored knowledge of the different areas of production. Just as the division of labour brings with it advantages, from a social point of view new possibilities are also opened up in terms of providing weekend relief and holidays for those working in these seven-days-a-week work areas.

As with all long-term partnerships, trust is essential for success, along with an interest in each other, a clear delineation between people and their areas of work, common spiritual work and a joint responsibility towards the surrounding world. Community farms of this kind are generally held within a charitable trust. This is a helpful starting point if we wish to solve the issue of the ownership of the means of production in agriculture, which is a very strong impulse in the biodynamic movement.

Farm and market garden partnerships

This is a form that is particularly common in France. Within the context of a mixed farm, a vegetable-growing enterprise is set up that is independent both financially and in terms of personnel. For the market garden this has many advantages. The manure, large machinery and designated fields within the crop rotation are provided by the farm, which in turn benefits from being closer to customers and many temporary workers who are employed during the intensive growing season. The high degree of independence also allows for better and more transparent ways of working together long term. In France there are several specific legal forms for agriculture that make this possible, encouraging specialisation within larger farm businesses.

Garden parks and grounds

In the anthroposophical movement there are many homes, conference centres, clinics and schools that have a large estate or campus to manage. This situation has given rise to a particular form of 'garden park'. This is a retreat or recreational space containing all kinds of features, especially trees, shrubs and hedges, meadows and lawns, water features and moist biotopes, paths and resting places interspersed with productive areas for growing vegetables, flowers, herbs, medicinal plants, soft fruit, orchards and composting sites. In many places, animals are also included in this parkland organism. One example is the 10-hectare (25-acre) estate of the Goetheanum in Dornach, Switzerland, on which cows are kept. During the grazing season the pastures are grazed by a herd of five or six cows with a bull and calves of the small *Rhätisches Grauvieh* breed. In winter they are housed with a neighbouring biodynamic farm. The cows thrive and bring more interest and a soul quality to the estate, the grounds receive manure, and walkers can occasionally witness the birth of a calf.

Figure 3.6. The Goetheanum estate in Dornach, Switzerland.

Growing wine in organic landscapes

When a vineyard first converts to biodynamics, it consists solely of vine rows and is usually far removed from being an agricultural organism. Through the biodynamic impulse, however, sensitive wine growers become aware of the surrounding nature and begin to cultivate the landscape. A copse of trees is planted as an island in the vineyard, offering nesting places to birds, and, if conditions allow, animals such as sheep, goats, cows and horses are also brought to the vineyard.

There are now biodynamic vineyards in the UK, France, Italy, Austria, Germany, Switzerland, Quebec, Chile, Argentina, South Africa, New Zealand, the USA and Australia. Alongside their vine rows they have developed into large and very diverse organic landscapes, many of which are open to visitors. Biodynamic vineyards are highly refined systems of production, with the surrounding landscape contributing to the fertility and ripening of the grapes. This results in first-class biodynamic wines that repeatedly win prizes.

Between the farm and the marketplace

From the beginning, economic questions have accompanied the biodynamic movement. Farmers are fully engaged with the land side of the economy, but they also need to manage the economic relationships they encounter in the marketplace. This can often be a challenging task. One practice that aims to humanise these economic relationships is associative economics. In this approach, representatives of the different elements of the farm-and-market economy – farmers, workers, distributors, consumers – form an association to set their priorities.

The first step to forming an association involves the different economic partners discussing the relevant economic processes that they share. The second step is to then assess the market. Are there too many or too few products? Are they properly priced for the market? How is demand likely to develop? Further questions can include the conditions and rights of workers and their ability to contribute to the development of the workplace. The third step is to develop the economic parameters (quality, quantity, price). In doing so all parties act within their own areas of expertise but with a shared understanding of the overall situation. The same economic rules are applied in an association as elsewhere, but the essential difference is that they don't play out in an anonymous or half-hidden way. They are deliberate, transparent and take place in partnership.

There is no fixed model for an association. They can be as small as one farm or as large as a national network; they can focus on a single product or on the total economic activity or an enterprise; they can be more concerned with the circulation of goods or with the provision of credit. In each case an association involves making a commitment to the group without losing independence.

Community Supported Agriculture (CSA)

The model of Community Supported Agriculture (CSA) was pioneered by Trauger Groh and his wife, Alice, on a farm in northern Germany, which they ran in partnership with two other families. When Groh emigrated to America, he set up a similar enterprise on a community farm in New England. The success of the project led to CSA initiatives appearing all over the country. Groh set down his experiences and gave practical advice in setting up CSAs in his two books, *Farms of Tomorrow* and *Farms of Tomorrow Revisited*.

In a CSA, a network of individuals and families connect themselves with a farm to form a supportive community. The farm then produces to meet the needs of that community according to an agreed plan, and the community agrees to meet the costs incurred in terms of wages, seeds, distribution and so on, regardless of the actual harvest. It is a localised association in which demand and supply are mutually determined and the price is based not on the goods but on a share of the harvest or the costs. The risks of a bad harvest as well as the benefits of a good harvest are shared in the community. Cooperation between producer and consumer is determined not by fixed prices, but by considering the needs of everyone involved.

The annual budget of a farm can be taken on by a consumer association, and in return products from the farm are distributed to the supporters. The association might be a regional one, it can even be a global one. An example of the latter is Teikei Coffee. Members from anywhere in the world buy a coffee subscription that supports farmers growing coffee beans in Oaxaca, Mexico. This provides the farming community with financial security and allows them to farm in a sustainable manner. Members then receive the coffee grown by the farmers.[10]

Other movements that connect producers and consumers directly also exist in Japan, Thailand and Korea.

Regional networks and link ups

A market that is protected by state tariffs in Switzerland offers a good example of associative work. The organic and Demeter market operates according to associative principles in that round-table discussions take place about each product category. Representatives from the two largest retail chains in Switzerland, Migros and the Co-op, along with local food processors and a representative for independent farmers all sit at the same table. The discussions are chaired by Bio Suisse and Demeter. Demand is considered and production is planned, the pricing structure is agreed and some measures for moderating it are put in place. The disastrous consequences for producers of the non-coordination of quantity, quality and price – the so-called free market – can be seen for example in the conventional industrial milk market, where rising costs mean dairy farmers are struggling to cover their expenses.[11]

Figure 3.7. A supermarket with organic products from Dottenfelderhof.

Organic sales and marketing

For a long time, there were no marketing structures for organic and Demeter products. As a result, the pioneers of biodynamic farming found themselves thrust into marketing roles. From the market stall an organic shop evolved and from it a wholesaler who in turn became an importer. In this way the developing trade in organic products became an integral part of the organic movement. It was not uncommon for these enterprising people to come from the Demeter scene for they had found inspiration in the course Rudolf Steiner gave on the world economy two years before he gave the Agriculture Course.

A second phase saw retail chains being established with wholesalers developing franchises. In 1984 Alnatura was founded by Götz Rehm, who was heavily influenced by Steiner's ideas. The company distributed food and textiles that had been produced in a sustainable way, and it was through this that many Demeter and organic products came on to the market. In 1987 the first Alnatura organic supermarket was opened and today there are more than 100 such organic supermarkets. The most recent step has been to introduce Demeter products into more mainstream supermarket chains.

Figure 3.8. Demeter-certified products in the supermarket.

Migros and the Swiss Co-op

In 1981, the Swiss Association for Biodynamic Agriculture helped to found Bio Suisse, an umbrella organisation for Swiss organic producers and the leading organic organisation in Switzerland. More than 7,500 Swiss farmers and gardeners and over 1,100 processing businesses sell products using its 'Knospe' label, one of the strictest organic food labels in the world. Food produced under Knospe is grown sustainably and processed with care. Between them these organic farmers manage more than 172,000 hectares (425,000 acres) of land amounting to 16.5% of the entire productive land area of Switzerland (in 2021). In 2014 Bio Suisse recorded a turnover of 2.2 billion francs ($2.6 billion), in 2020 it was 3.8 billion ($4.5 billion) and in 2021 more than 4 billion ($4.75 billion). In Switzerland organic products make up nearly 11% of the market.

Co-op and Migros are the two largest purchasers of organic products in Switzerland. Since 1993, Co-op Naturaplan has been selling organic food in its supermarkets. Organic products are omnipresent in the Co-op supermarkets, with over 4,800 organic products available in the Co-op producing a turnover in 2021 of 2.1 billion francs ($2.5 billion). These are not the Co-op's own products but come from external suppliers.

The Migros range includes a total of 3,200 food products from Migros Bio, Alnatura and Demeter. Organic products in Migros make up 12% of its turnover.

Organic, biodynamic or ecological?

The term for organic agriculture in Germany is 'ecological agriculture'; in Switzerland and Austria it is 'biological agriculture'. By contrast, 'biodynamic agriculture' is used throughout the world to describe the form of agriculture bearing the Demeter label.

In the 1940s, the organic-biological movement in Switzerland came into being to distinguish itself from the anthroposophically inspired biodynamic agriculture. Nevertheless, since then the two have worked closely together. This manifested itself in a concrete way with the founding of the Research Institute of Organic Agriculture (FiBL) in 1973 and Bio Suisse in 1981.

In Germany until the beginning of the 1970s, the only organic-agriculture organisation consisted of biodynamic farmers. Gradually, however, additional organisations were founded, such as Bioland in 1971 and Naturland in 1982. At the end of the 1980s the various organic organisations and the Research Circle for Biodynamic Agriculture came together to form the Organic Agriculture Co-operative (AGÖL) as an umbrella organisation to address common concerns with the state and EU regarding quality guarantees for their products. Since 2002 this has been known as the Organic Food Federation.

Scientific studies

Research has been part of biodynamic agriculture from the beginning. Immediately after the Agriculture Course, the farmers who had attended it set up comparative trials on their fields. They established plots where they applied the preparations and plots where they didn't to act as a control.

In the second half of the twentieth century a number of independent research institutes for biodynamic agriculture were established, including the Biodynamic Research Institute in Germany in 1950, the Nordic Research Institute in Sweden in 1956, the Louis Bolk Institute in Holland in 1976, and the Michael Fields Agricultural Research Institute in America in 1984. But the first institute of this kind was set up by the Natural Science Section at the Goetheanum in Dornach, Switzerland in 1921.

Scientific studies and research into organic and biodynamic agriculture have since been carried out at numerous agricultural universities. The first dissertations on biodynamic themes were written in 1973 at the University of Giessen in Germany, and the first explicitly biodynamic postgraduate thesis, by Dr Hartmut Spiess on rhythmical research and chrono-biological sowing trials, was accepted by the University of Kassel, also in Germany, in 1994.

A high-profile example of an experiment comparing growing systems, is the DOK trials run by the Research Institute of Organic Agriculture (FiBL) and the state research centre Agrosope (see the following section for a more detailed description). The trials began in 1978 and compared three farming systems: biodynamic (D), bioorganic (O) and conventional agriculture (K, from the German *konventionell*). These have found that for biodynamic soils all the important parameters needed for soil fertility are present. In a joint US and Spanish project to measure fungus colonies in the soil, 350 soil samples taken from vineyards in the United States and Spain showed that in biodynamic soils colonies developed a more compact structure and were therefore less affected by environmental changes than was the case with conventional soil cultivation.[12]

A French research team also found that the effects of environmental stress were less pronounced on biodynamic farms. The researchers were able to show that biodynamic vines reacted to drought conditions with greater genetic activation (which plays an important role in cell differentiation, growth and response to stimuli) and were therefore more adaptable than conventionally grown vines. In addition, the average weight of yields in conventional cultivation showed significant decline in years of drought while biodynamic yields remained stable.[13]

More recently there have been a series of trials into the effectiveness of the horn manure and horn silica spray preparations. Squashes treated with horn manure resulted in a significant increase in both the total and marketable yield.

With horn silica treatment, the marketable yield, as well as the content of macro-nutrients, total carotenoids, individual carotenoids and antioxidants, was significantly increased. The two preparations together had a significant effect on total and marketable yields, and on the net photosynthesising capacity, dry matter content and the content of total and single carotenoids. It can be taken from these results that horn silica in particular promotes the ripening of fruit. This offers a way of proving the high quality of Demeter certified biodynamic food scientifically.

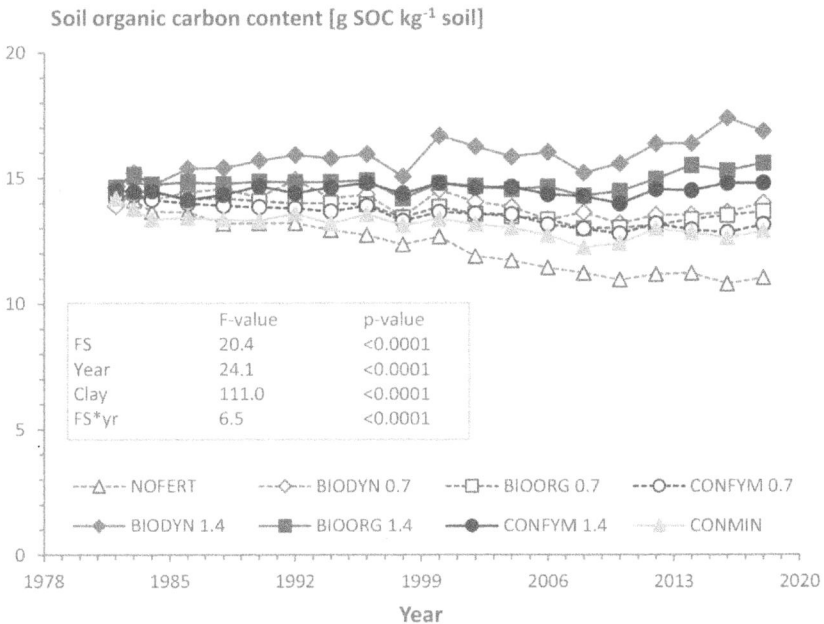

Soil organic carbon content [g SOC kg⁻¹ soil]

	F-value	p-value
FS	20.4	<0.0001
Year	24.1	<0.0001
Clay	111.0	<0.0001
FS*yr	6.5	<0.0001

--△-- NOFERT   --◇-- BIODYN 0.7   --□-- BIOORG 0.7   --○-- CONFYM 0.7

—◆— BIODYN 1.4   —■— BIOORG 1.4   —●— CONFYM 1.4   —▲— CONMIN

Figure 3.9. Amount of organic carbon in the soil (gram per kilogram) over a period of 42 years in a comparison between BIODYN (biodynamic), BIOORG (Swiss bio-organic), CONFYM (conventional with manure), CONMIN (conventional with artificials).

Comparison of biodynamic, organic and conventional growing systems

Results from the long-term DOK trials have been particularly illuminating. The field trials take place on loess soil – a pale, loose type of soil that is formed by wind or glaciers and is found in Europe, North America and Asia. The soil is continually assessed and adjusted by groups of professionals to ensure that conditions for each of the three growing systems are as closely aligned as possible. Over the years a clear divergence between the systems has emerged in the scientific data, which has formed the basis of 150 peer-reviewed articles. On average, biodynamic yields are 20% greater than those grown using organic methods.

It is important to note that throughout the trials the difference in yield between organic and conventional agriculture has been constant for over forty years. The organic system maintains soil fertility and remains productive while using around 50% less fertiliser and energy inputs. This means both the biodynamic and organic systems are more efficient and less damaging to the environment. The main differences between the biodynamic and organic systems were most noticeable with two crops: potato and wheat. Organic potato yields were 15% higher than biodynamic yields, which can be traced back to the use of copper sprays. In the case of wheat, however, biodynamic yields were 85% higher than organic ones. A significant reason for this is likely to be the use of the biodynamic-bred wheat varieties of pioneer Peter Kunz, the founder of the plant-breeding organisation Getreidezezuchtüng Peter Kunz (GZPK).

Figure 3.10. After heavy rain the biodynamic plot (bottom) is significantly less waterlogged than the conventional one (top). This is largely due to the many earthworms, whose tunnels allow water to trickle down.

In the organic and biodynamic sectors respectively, 95% and 100% fewer toxic substances were found than in the conventional one. This is important not only for the insect world but also for our attempts to provide uncontaminated food and clean drinking water. We can see the differences with our own eyes as well. After heavy rain organic plots are noticeably less compacted. The soil structure of the organic variants is more developed and stable. There are more earthworms on organic fields and water can trickle through their burrows. Overall, some 30% more soil life was found in organic soils and 60% more in biodynamic soils compared with conventional plots.

Measurements in the DOK trials show that the organic plots produce 36% fewer greenhouse-gas emissions and the biodynamic plots 61% fewer. This is due to reduced nitrogen inputs, better soil structure, more stable pH values and colonies of microbes that can convert nitrous oxide into harmless nitrogen gas. The biodynamic system is particularly climate friendly since it is better able to keep carbon in the soil in the form of humus.

In summary, the scientific results of the DOK trials provide definitive confirmation of the goals and values of the biodynamic approach. With all the parameters for a healthy and sustainable agriculture – soil fertility, biodiversity, a positive effect on climate – the biodynamic plots produced high values. The contribution made to the ecosystem by the biodynamic approach has been found to be significantly higher than that of other cultivation systems. Yields have remained at 80% of conventional yields for decades, demonstrating that there is security of yield over the long term. The claim made by critics of organic methods that the soil will become impoverished due to the lack of nutrients provided by chemical fertilisers is therefore wrong. The opposite is in fact true. If the energy and environmental effects are also considered, we can rightly speak of the organic and biodynamic variants as being more efficient. Conventional variants require large inputs, while inputs in the organic variants are almost nil. This is especially true of the biodynamic variant according to the findings of the DOK trials.

Research into organic viticulture

At the Institute for Traditional and Organic Viticulture, which is connected to the Geisenheim University in Germany, there is an ongoing research project called INBIODYN.[14] Begun in 2006, it is a scientific, comparative field trial, modelled on the DOK trials, that seeks to investigate the effects of the conventional, organic and biodynamic systems on vines and vineyards. Regarding microbial activity in the soil and soil biodiversity, as well as the quality of the wine, the results from these three approaches were striking. Organic and biodynamic cultivation indicated a higher level of enzyme activity, more fungal and bacterial activity and a greater variety of bacteria.

As a result of this research project, Geisenheim University is being promoted as an 'Innovative University' by the federal government and Hessen state for the next five years with a grant of more than €2 million.

The organic wine movement in France

Wine makers can discern the influence of biodynamic cultivation in the smell and taste of the wine. The flavour conveys the context from which the wine arose – the so-called *terroir*, which refers to the soil, the climatic conditions of the year and sometimes even the personality of the grower. Philippe Faure-Brac, the world's best wine connoisseur of 1992, describing his tasting experience said:

> Biodynamic wine appears to have more acidity and depth. The most important difference is the wine's balance of flavour. The mineral quality is more strongly developed, the personality of the wine comes more uniquely to expression.[15]

Olivier Poussier, best connoisseur of the world in 2000, declared: 'My most wonderful tasting experience has been wine of biodynamic origin.'[16]

Since the 1990s there has been a thriving organic wine movement in France. There are now 609 certified biodynamic vineyards, 469 with the Demeter label, which has been used for biodynamic wine since 2009, and, as of 2021, 140 with the Biodyvin label.[17] Demeter wine makers frequently win awards and, as leaders in the wine business, they are among the top wine makers of the world (take for example, Gault et Millau, Gilbert et Gaillard and Bettane Desseauve to name a few).

In the February 2021 edition of *Revue du Vin de France*, more than a hundred organic wines were profiled.

> *No backward-looking orientation:*
> "In speaking here from an anthroposophical point of view, it is really important not to revert to the old instincts but to find, out of a deeper spiritual insight, that which instincts ... can provide to an even lesser degree. For this it is necessary that we devote ourselves to extending our studies of the life of plants, or animals and also of the life of the Earth itself to encompass cosmic dimensions."
>
> *Lecture One, June 7, 1924*[18]

Biodynamic agriculture and the traditions of Indigenous peoples

Since 1925 an annual agriculture conference has taken place at the Goetheanum. It is a meeting place for the biodynamic movement where farmers, processors, traders, researchers, advisors, students and apprentices from anywhere in the world, as well as those who are interested in the subject and want to learn more, can exchange experiences and deepen their knowledge. The conference is now transmitted online in seven languages to around 1,000 participants from 50 countries.

In February 2020, just before the Covid lockdown, representatives from indigenous communities across the world were invited to share their experiences. The title of the conference that year was 'Paths to the Spirit in Agriculture', and it was focused on exploring the connections that exist between the biodynamic agriculture born in Europe and the spiritual understanding and cosmological traditions of indigenous peoples. Many agricultural practices that are known today have their origin in indigenous agriculture where the unity between human beings and nature has been held in high regard.

Agriculture as a cultural impulse

For millennia, people who lived an agricultural existence had to battle the forces of nature and were often subject to their land being owned or taken away by others. We can feel great respect and gratitude for the people of pre-industrial times who, through their insight and dedication to nature, created the cultivated landscape. They domesticated animals, bred crop plants and produced fertile soil. Even today we can experience the after-effects of diverse agricultural impulses from ancient times – for instance, the intimate partnership with nature of indigenous farmers, the interaction with holy cows in India, and the cultivation of wheat seeds over many generations that began in the Middle East. In Central America, the Mayans and other Mesoamerican peoples grew maize, beans and squashes together. The maize served as support for the beans, the beans brought nitrogen into the soil to be used by the other plants, and the leaves of the squash plants covered the soil, suppressed weeds and kept the soil moist. The Milpa system is based on this, along with other variations.

During this course of cultural development, we discover again and again that there have been phases of self-determination, collective creativity and responsibility for taking agriculture forward. Periods of healthy agricultural development were always culturally inspired while cultural development was inspired and carried by agriculture. Humanity and the earth, culture and nature – our being human and becoming human – have an existential and co-dependent relationship to one another. The perspective of the history of consciousness in relation to the evolution of agriculture can give us an idea of where we – the whole of humanity – are at present and what evolutionary challenges we face today and in the near future.

Future perspectives

To understand agriculture in its entirety ever more deeply and to give it form as an organism and as an individuality will remain a challenging task long into the future. Many of the social, political and even global problems of agriculture take on a more fruitful character when considered from this angle.

This applies to economic questions, among others. Today agriculture is perceived as a trade or an industry, and as such it must continually exploit nature and society. It is hardly able to sustain itself and therefore must be subsidised by the state. However, if it is understood instead as a form of primary production and is organised and managed accordingly, it can provide a healthy counterbalance to the resource-hungry activities of industry. If biodynamic and organic agriculture can demonstrate these interdependencies more clearly, then agriculture will take up the healing role it once had within human society and which it has largely lost today. This will not be achieved by returning to traditional relationships but by courageously taking up new forms of cooperation and ways of working with nature.

4. The Biodynamic Movement 100 Years On

Biodynamic agriculture across the world

One hundred years after Rudolf Steiner gave the Agriculture Course, biodynamics is practised all over the world. From its early beginnings in Europe, emigrating Europeans carried it to the USA, South Africa and Australia. It has since arrived in North Africa, India, China and Latin America, and is currently taking root in many countries in South East Asia and other parts of Africa. For those committed to biodynamics it is more than just another agricultural method. It is an opportunity for them to experiment with different, more sustainable and mutually beneficial ways of living and working together. To this extent biodynamics can be seen as a cultural impulse at work in the realm of farming.

Demeter

Demeter has existed in Germany since 1928, when it was established by Georg Michaelis, who had been part of the group that purchased Marienhöhe farm. It was created to identify biodynamic products in the marketplace and indicate that they adhered to certain standards of cultivation and production. It is not only about what is done but also *how* it's done. Since 1932 Demeter has also been a registered trademark. In Switzerland, the trademark was administered by a biodynamic association until 1997, when the Demeter Federation was founded as a certification body. Similar developments have taken place in many European countries. 1997 saw the founding of Demeter International, a coming together of all the countries that use the symbol. Its purpose was to manage and coordinate marketing policy in a cooperative way. This way of developing and collectively managing a trademark is unique and clearly expresses the fact that the trademark should be understood not as a business tool but rather as a bona fide expression of biodynamic quality and a bridge to bring producers, food processors, traders and consumers together.

Figure 4.1. The Demeter logo

Demeter certification is based on the requirement to meet standards set out for production and processing. These common rules are determined democratically on a national and international level. Proposals to alter standards come from new research results, practical challenges and sometimes also marketing interests. Delegates from all member countries have a vote to determine what should apply.

The annual inspections, resulting certification and the regulated application of the label to the products takes time and carries costs. The Demeter label opens the door to the highly differentiated marketplace, especially in Europe and North America. The turnover in Demeter products is continuously growing.

At the end of 2021 there were 7,000 certified farms across the world, covering a total area of 226,000 hectares (558,450 acres). To this are added 1,100 food-processing businesses and 560 trading companies with a Demeter contract. There are Demeter-certified businesses in 60 countries. In the specialised wine-growing sector, which has developed strongly over the last 15 years, there are 1,300 business in 22 countries on a total of 22,000 hectares (54,350 acres).

By no means are all biodynamic farms Demeter certified. For the local marketing of products certification is not worthwhile, especially in countries where Demeter is unknown. In some countries an accreditation through the Participative Guarantee System (PGS) is being developed. Numerous projects involving many thousands of small farmers engaged, for example, in the growing of cotton market their products under an organic label. Their farming, however, is fully biodynamic. It is estimated that there are about 100,000 farmers worldwide who know about and apply many biodynamic principles.

Biodynamic agriculture amidst world challenges

Biodynamics has an important contribution to make to the many challenges facing agriculture and food production today. Apart from a necessary focus on agriculture and food production at the regional level, a wider ecological consciousness is also needed. Climate change, soil erosion and the loss of biodiversity are already the cause of starvation, and this will only get worse in the future. The strength of biodynamic and organic agriculture lies in the fact that, although yields are 20% lower than those of conventional methods, they nevertheless have a positive impact on the climate and ecology. Biodynamic and organic systems *can* feed the world, but this alone is not enough. It will be necessary to moderate meat consumption and reduce food waste. For instance, it is estimated that nearly half the global wheat crop is used for fattening animals, and then there is 'wheat for heat' used in the many bio-gas units.

In 2008, the World Development Report formulated a key strategy for creating a sustainable agricultural system of the future, one that was regional, ecological, multi-functional and experience-based.

Chemical agriculture, however, also lays claim to the future. The CEO of the multinational pharmaceutical and chemical company Bayer, which has incorporated Monsanto as the producer of the systemic pesticide Roundup, has proposed using advanced technology to solve the current grain shortage caused by the war in Ukraine and the threat of starvation facing millions of people. This would involve using genetically modified wheat to fix nitrogen in the soil and the development of smart farming or digital agriculture, which among other things means introducing artificial intelligence into agriculture. A training programme for a hundred million small farmers has also been suggested. This prospect of a so-called regenerative agriculture sounds like a rehashed version of the 'green revolution' of the 1960s whose promised benefits have never yet been realised. In 2022, the CEO of Syngenta blamed organic agriculture for starvation in Africa because it only produces half the yield per hectare compared with chemical agriculture. But as the long-term DOK trials of FiBL have shown, yields are 80% of conventional yields, not 50%, with far lower inputs required in terms of energy and a far better ecological balance sheet.

> "It will come to be seen that mineral fertilisers must in time completely cease; for the effect of every kind of mineral fertiliser is that after some time what is grown on fields where such fertilisers have been used loses its nutritive value. This is a wholly universal law. Moreover, if what I have indicated is taken up, it will not be necessary to manure more often than every three years. You may perhaps not need to manure more often than every four to six years, and you will be able to do without artificial fertilisers altogether. If you apply the methods I have indicated, you will dispense with such fertilisers if only because it will be much cheaper to do so. Artificial fertilisers will no longer be needed; they will disappear again. People judge everything in accordance with much too short spans of time."
>
> *Lecture Six (Discussion), June 14, 1924*[1]

An end to chemical fertiliser use

Biodynamic agriculture has been practised for 100 years without artificial nitrogen fertilisers, pesticides and herbicides, and without GM technology. It has also been shown that along with other organic and ecological systems, it is possible to grow high-quality produce using a sustainable approach to the cultivation of soil, plants and animals. The idea that in order to feed ourselves we must pump nature full of poisons has long been transcended. The grave harm done to the soils, waters and atmosphere of the earth makes it clear that we need to rethink and change course. The biodynamic approach to agriculture and nutrition has shown it can be done, as the following examples make clear.

Sekem, Egypt

The Sekem initiative, which takes its name from the hieroglyphic transcription meaning Vitality of the Sun, was founded by Ibrahim Abouleish in 1979 on 70 hectares (170 acres) of desert land, which he inherited on the fringes of the Nile Delta about half an hour's drive from Cairo. With access to the land's own spring water for irrigation, Abouleish began by planting hedges, growing clover, keeping cattle and cultivating many different herbs.

Since then, many contract farms and new independent initiatives have been added. Among them is Wahat, a farm in the White Desert 300 kilometres (185 miles) south-west of Cairo, which is being intensively developed. The 1,000 hectares

(2,500 acres) is being cultivated using groundwater that is applied sparingly via large circular irrigation systems. The first stages have been completed and the compost imported from the mother farm is gradually being replaced with material produced on site. A system of livestock management is being developed. Several different scientific research programmes are being conducted that monitor Wahat's developing biodiversity and investigate soil bacteria. More and more people are living on the site, with houses, a hotel, a restaurant, a small school and a theatre being built to accommodate the growing community.

Figure 4.2. Sekem, Egypt

As an enterprise, Sekem consists of six companies with nearly 1,900 workers. Fresh and processed food, medicines, textiles and much else is produced and sold both in Egypt and around the world. The workers are trained, have social insurance and access to medical support. There are kindergartens and schools for the children, craft training courses for young people and integrated care for those with special needs. Health provision has even been extended to include the surrounding villages.

Biodynamic agriculture has made it possible to transform the desert into fertile land. The soil is typically categorised as 'aridisol', meaning it has little water and extremely low levels of organic matter. Soil fertility is built up using large amounts of compost made from plant and animal waste. Thanks to the research work taking place on site, a biological pest-control system was also developed to produce organic cotton. As a result, the Egyptian state was able to end the routine spraying of cotton fields over many hundreds of thousands of hectares and save on 35,000 tons of pesticides.

To promote biodynamic agriculture and produce the raw materials required by the Sekem businesses, the Egyptian Biodynamic Association (EBDA) was founded. This association has helped to convert 21 farms out of more than 170 to biodynamic methods with Demeter certification across 5,000 hectares (12,350 acres) of land. In 2003, Ibrahim Abouleish received the Right Livelihood Award, often called the Alternative Nobel Prize. He was recognised for the development of a twenty-first-century business model in which economic success is integrated within the social and cultural development of society, which in turn advances the 'economy of love'.

Sekem in numbers

⚶ A total of 1,840 people work for the Sekem holding, across agriculture, construction, hospitality, sustainability and certification.

⚶ The Sekem holding, consisting of the home farm and four other farms in different parts of Egypt, manages more than 2,500 hectares (6,175 acres) biodynamically.

⚶ Sekem also works with 26,000 small farmers who manage a total of 20,000 hectares (49,420 acres) biodynamically or organically.

⚶ The Sekem Medical Centre provides medical support not only for those working for Sekem but also for the 30,000 inhabitants of neighbouring villages. The income in 2021 was €3 million. Of this, 50% came from the Sekem holding, 25% from donations and 25% from grants, primarily through research projects with the EU.

⚶ The Heliopolis University for Sustainable Development was established by Sekem in 2009 to introduce the principles of sustainable development to students and the wider Egyptian community. Since 2012 it has consisted of five faculties: pharmaceutical, business and economics, engineering, physiotherapy, and organic agriculture. It has 287 academic staff and 2,825 students. In 2021 it had an income of €2 million, with 70% coming from student fees, 20% from donations for scholarships and 10% from grant funding.

Capão Alto das Criúvas, Rio Grande do Sul, Brazil

The estate of Capão Alto das Criúvas is just over 550 hectares (1,360 acres). It is run by João Volkmann and has been under biodynamic management since 1983. Volkmann's father, who was familiar with anthroposophy, bought the farm in 1954 and started work on conserving the land. Between 1974 and 1983, however, the farm was rented out to a conventional farmer and became severely polluted with agrochemicals. When João Volkmann took over the farm he decided to focus on growing rice. By applying the biodynamic preparations and controlling the water table, it was possible to bring the natural flora and fauna of the complex rice ecosystem into harmony, increase soil fertility and in short time eliminate the smell of methane in the rice fields. Two hundred and fifty varieties of birds have now been observed there.[2]

Figure 4.3. Buffalo grazing in Rio Grande do Sul, Brazil.

Since 1999 the rice grown at Capão Alto das Criúvas has been Demeter certified. As well as rice, João Volkmann also grows beans, maize, manioc and many other kinds of vegetable for the farmers, land workers and their families. More than 100 cattle, some 100 water buffalo and a few horses are at home on the land and graze the rice stubbles in winter.

Like many who manage biodynamic farms, Volkmann offers introductory courses on biodynamics and courses on making the preparations.[3]

Urban gardens in Rosario, Argentina

In the Argentine city of Rosario, a consortium of local organisations and city authorities have brought nearly 1,000 hectares (2,500 acres) of urban land back into cultivation as part of an organic and biodynamic urban garden (*heurtos urbanos*) initiative. The Rosario Urban Agriculture Project is an extensive long-term project that works with the inhabitants of the town and uses agroecology, permaculture, organic and conventional agriculture to bring about the transformation of disused urban spaces. Gardeners have converted rubbish dumps into gardens and brought the town into flower. A centre for organic agriculture has been established in Rosario and offers courses on biodynamic composting techniques and the preparations, as well as on the use of an astronomical sowing calendar. The city's weekly organic market is the first of its kind in Argentina.

Figure 4.4. An urban garden in Rosario, Argentina.

In 2004, the city farm in Rosario received the UN-Habitat Dubai Award for one of the ten best initiatives in the world to protect the environment and reduce poverty, and in 2014 the project featured in the French documentary *Growth – What Next?* by Marie-Monique Robin. In 2021 the Rosario Urban Agriculture Project won first prize from the WRI Ross Center Prize for Cities, which 'recognises trailblazing projects and initiatives for their contribution to inclusive and sustainable urban transformation'.[4]

Antonio Luis Lattuca in Rosario

One of the leading brains and builders of this extensive urban agriculture and garden project in the million-strong city of Rosario is the agricultural engineer and ecologist Antonio

Luis Lattuca. For Lattuca, the Urban Agriculture Programme (UAP) offered a constructive solution to Argentina's disastrous economic crisis of 2001, in which half the population of Rosario fell into poverty. Lattuca became the link between the city authority and the network of local organisations of agroecology, permaculture and biodynamic farmers.[5]

Ninety per cent of Argentina's population lives in cities. According to Lattuca:

> People have grown cold, profit-orientated, mechanistic and lifeless. It is vital that people are inspired to create a new life and vigour and find quality of life in a centre of civilization once more.[6]

As a member of the Argentine Association for Biodynamic Agriculture (AABDA), Lattuca knows from many years of experience that it is up to the farmers to make the soil in and around the cities fertile again without chemicals and by using compost, organic manure and biodynamic preparations. As he describes it:

> Today in this current crisis we are called upon more than ever to cultivate the soil, make new furrows and sow seeds. It is a great opportunity and challenge to cultivate all this abandoned industrial land in and around our cities. Our towns are calling upon us to cultivate them.[7]

Alliance for the Earth, Spain

The vineyard of Gramona in Penedès, 30 kilometres (18 miles) from Barcelona, has been biodynamic since 2016. It is a model farm of 100 hectares (250 acres) that integrates sheep and horses alongside the vines. Its sparkling wines regularly meet and exceed the 90-point level in the 100-point system, making them among the best wines in the world for connoisseurs. Together with twelve other neighbouring farms, Gramona founded the Alliance for the Earth (Alianzas por la Tierra), which has transformed over 300 hectares (750 acres) of a wine-growing monoculture into a diverse biodynamic landscape. It is a good example of how cooperation rather than competition opens the way to the future.[8]

Figure 4.5. The Gramona vineyard

Binita Shah, compost queen

Binita Shah is a farmer and landowner in the village of Supi in Uttarakhand in northern India. Her farm of 5 hectares (12 acres) lies 2,400 metres (7,900 feet) above sea level in the Himalayan foothills. This region receives an enormous amount of rain and 30 tons per hectare of topsoil are lost each year through flooding. The terraces are often damaged by the monsoon rains and each winter have to be repaired.[9]

Binita Shah has played a leading role in the spread of biodynamic agriculture in India. In 1998 she founded her consultancy firms Supa Biotech (the first part of its name stands for Steiner's Universal Philosophy of Agriculture), which is a leading producer of biodynamic preparations. In 2000 Binita Shah mounted an initiative in Uttarakhand that led to 100,000 compost heaps being made using the biodynamic preparations. Since then, over a million compost heaps have been made as part of this project run by the Indian government to strengthen the independence of 50,000 small farmers.

Figure 4.6. Compost heaps created by Binita Shah.

Supa Biotech's sister organisation, SARG (Steiner's Agricultural Research Group), assists farmers in converting to organic and biodynamic agriculture. In 2012, SARG received the prestigious Krishi Bhushan Award from the Maharashtra state government for its 'exceptional contribution in promoting organic agriculture'. Supa Biotech is currently involved in 29 growing trials and yield assessments as part of a long-term working relationship with regional agricultural universities.

Binita Shah's initiative is an example from northern India. There are also other large projects in central and southern India, where the viability of many family farms, and the health of the families working them, is being improved through biodynamic practices. These well-organised initiatives are often built around an export commodity such as tea or coffee.

The Barrel or Cow Pat Preparation (CPP) is widely used in India. Originally developed in Germany by Maria Thun in the 1970s, CPP helps to strengthen plants and increase their resistance to pests. It is also used for treating seeds. CPP is made in a pit lined with bricks and the compost preparations are added to a mixture of fresh manure from lactating cows, ground eggshells, basalt meal and unrefined sugar. Using only a small amount of the preparations, 40 kilograms (90 pounds) of CPP can be produced and applied as a combined preparation. There are also forms of manure concentrate produced according to the more traditional forms of Ayurvedic agriculture. Planting calendars are also known in India. Because there is a great reverence for the cow in Hindu traditions, the biodynamic approach is more readily accepted.

Sowing the future

Every farmer and gardener needs seeds each year, but the seed market today is fiercely contested. It is dominated by a few large corporations that use hybridisation and gene technology to produce different varieties of seed, which they then patent. This prevents others from using, producing or distributing those seeds. But it also interferes with seed saving. This is a common practice in agriculture whereby farmers and growers collect seed from a successful crop and store it for future use. This helps them cultivate seed diversity and produce varieties that are more resilient and adaptable to the local environment. Seed patenting effectively criminalises this practice by centralising seed ownership among the main agrochemical corporations. As a result, individual farmers and growers can find themselves owing money to these corporations for seeds they have gathered from their own crops.

It is against this background that an independent biodynamic plant-breeding programme has come about. From very simple beginnings a whole network of plant breeders, seed producers and marketing organisations has developed. A range of biodynamic cereals along with the first varieties of arable legumes and maize, and almost all the commonly grown vegetable varieties that have been bred biodynamically and entered in the seed register, are available. The first candidate varieties of apple are in the process of being registered. As a result, nearly all the seeds used on farms or gardens in central European countries today come from this network.

Seed initiatives like this one that have developed within the biodynamic movement demonstrate how much potential there is in the Agriculture Course.

Dottenfelderhof: plant breeding and research

Dottenfelderhof is a biodynamic farm in Bad Vilbel, Germany, that combines basic scientific research with agricultural practice. The farm also includes an agricultural training programme. In 2021, the scientific director of Plant Breeding & Research, Dr Hartmut Spiess, received the Federal Order of Merit for his pioneering work with organic agriculture, in particular in the field of seed health and plant breeding. His research group is counted among the leading plant-breeding initiatives for organic cereals and vegetables, with a number of varieties now entered in the seed register. Among them are the first ever organic wheat varieties bred to be resistant to fungal diseases such as leaf rust and loose smut, and with an exceptionally good baking quality. Hartmut Spiess has since passed the running of the operation on to the next generation.

Figure 4.7. Dottenfelderhof farm community and training school in Bad Vilbel, Germany.

GZPK: Getreidezüchtung Peter Kunz

GZPK was started in Switzerland by Peter Kunz in 1984 as a one-man operation. Today it is a plant-breeding enterprise with fifteen employees. New varieties, especially of wheat, rye and triticale, are continually being released on the market with the result that 65% of the organic wheat grown in Switzerland now originates from GZPK varieties. It is very much the case that the varieties developed according to guidelines given by the Agriculture Course are gaining recognition far beyond the biodynamic movement. The greatest success is that of the Wiwa variety. It provides stable yields for the farmer, a greater output of flour for the miller, more loaves of bread per kilogram of flour for the baker, and exceptional nutritional value for the consumer.

GZPK is structured as a small charitable organisation. At the beginning the work was funded by private donations and foundation grants. For nearly twenty years there was a very supportive partnership with, and grants from, the Swiss Co-op. Since then, income from the variety licences has gradually increased and this makes it possible to finance the development of new varieties.

These two examples, Dottenfelderhof and GZPK, are representative of the many plant breeders who have been breeding and maintaining grains and field crops for decades and have provided farmers and food processors with a wide range of varieties. For many years, the plant-breeding fund Zukunftsstiftung Landwirtschaft (Foundation for the Future of Agriculture) has supported people who are dedicated to breeding cultivated plants. The trust supports pioneering projects that promote organic and sustainable land management (especially biodynamic agriculture), plant breeding and seed production.

In 2021, BioSaat GmbH was founded to promote the production of organically bred varieties of cereals and arable crops. Its aim is to make their varieties better known and facilitate their distribution in Germany, across Europe and throughout the world.

Vegetable seeds

In Germany, the biodynamic breeders of new vegetable varieties are part of the association Kultursaat. So far, over 100 new varieties have been developed and assessed by the Federal Seed Register of Germany. Bingenheimer Saatgut is the trading and production company for these and other varieties of vegetable seeds, and a comprehensive range is now available for organic growers and amateur gardeners. Bingenheimer Saatgut has grown rapidly and now has over 100 employees and more than 100,000 direct customers. In 2021, Kultursaat and Bingenheimer Saatgut received the Organic Farming Innovation Award for Science (OFIA).

In Switzerland, Sativa Rheinau is the largest and most prominent plant breeder, producer and trader of biodynamic seeds. New varieties from their own breeding programme are offered for sale each year, and they form partnerships with other seed producers – for example, ProSpecieRara, a non-profit charitable organisation that preserves old, local varieties.

New markets are actively sought, especially in France and Italy. The business is organised as a limited company, which allows for the active involvement of shareholders. Sativa Rheinau is a member of the business association Fintan, and the land it uses for its own plant breeding and seed production is integrated within the 120-hectare (300-acre) Demeter farm Gutsbetrieb Rheinau.

In 1998, Reinsaat was founded in Austria. It has fifty employees and now offers a comprehensive selection of seeds. Among them are many of Reinsaat's own varieties as well as those of the farms that partner with them.

Smaller initiatives produce and market biodynamic seeds in other countries. Plant-breeding skills are mastered, and the marketing is carried out professionally. An effort is being made to charge a levy on the turnover of organic food at the point of sale, thereby helping to fund plant-breeding activity.

Vandana Shiva: the campaign for patent-free seeds

In 1994, scientist and activist Vandana Shiva established the Navdanya initiative in India, a seed bank and organic farm run entirely by women. This was at a time when global corporations were seeking to legally secure control over seed production through patent protection, and she was determined to protect the diversity of seeds and the right of farmers to continue producing, breeding and sharing their seeds freely. Shiva's determination to challenge the toxic cartels was based on her belief in a future with organic agriculture, biodiversity and seed sovereignty:

We need a renewal of small farms, real farms with real people who are concerned for the land, life and the future, and who can produce a diversity of healthy, fresh, organic and genuine food for everyone.[10]

To date, over 3,000 varieties of rice have been preserved in Navdanya's seed bank, and on the Navdanya farm more than 2,000 varieties of crop are grown along with more than 150 species of tree. In 1993, just before she founded Navdanya, Vandana Shiva received the Right Livelihood Award for her contribution to developing policies supporting the position of women in society and ecology. In 2023 she was invited to address the annual Agriculture Conference at the Goetheanum in Dornach.

Across India, millions of farmers are converting to organic agriculture. In the state of Sikkim a law was passed in 2016 forbidding the use of pesticides, artificial fertilisers and GM technology and allowing only organic agriculture.[11]

Animal welfare in agriculture

Animals play an important role in biodynamic agriculture. In the Agriculture Course, Rudolf Steiner explained that the transformation of fodder within the animals' metabolism enabled a more or less closed cycle of substances to come about, turning the farm into a self-contained organism. This applies most of all to cows and bovine animals because the metabolism of ruminants is so highly developed. The use of farmyard manure leads to a continuously improving level of soil fertility, particularly through the development of site-adapted microbial life. If managed well, a unique farm biome develops that becomes visible throughout the farm in the health and resilience of the soil, animals and the crops grown.

In recent years, animal welfare has become a major issue. A great deal of pioneering work has been done in this area by Michael Rist, who held a professorship in livestock management and housing at the Swiss Technical College ETH Zürich. In the 1970s and 1980s he undertook research with his students into 'species appropriate livestock management', which in those days was a novelty. He described his approach to the research in this way:

> The lack of thought in agriculture has resulted in no longer considering the needs of livestock or the growing conditions our crops require, but only in asking how to maximise economic returns. By pursuing this short-term and purely economic objective there has arisen a necessity of applying environmentally harmful measures, something that is anathema to every farmer ... In the following contributions, examples of cattle, pig and poultry management will be given to show how little consideration is given to species appropriateness in conventional livestock systems and how a system that respects the nature of the species can be arrived at. In the process the economic aspects are then relegated to second place ... In this way consideration is given first of all to how livestock should be managed – that is, what is species appropriate – and only then about its economic realisation.[12]

Loose housing for horned cattle was developed on biodynamic farms around thirty years ago. This allowed the animals easy access to pasture, which has always been seen as important. The use of home-produced feed brings about a healthy balance between the land and livestock population, and the integration of animals within the farm organism has a stabilising and calming effect. This shows itself in the good health of the herd. New additions to the herd generally come from the farm's own young stock; the buying-in of animals is the exception rather than the rule.

Another aspect common to biodynamic farms is the diversity of domestic animals and the care given to the landscape that encourages the presence of a wide range of birds, insects, mammals and reptiles. The presence of animals leads to the farm and the landscape being permeated by a quality of soul. The entire animal kingdom is made up of specialists. Each species of animal brings a unique quality to expression in its surroundings and hence leaves a mark on its habitat. Just think how different the smell of a pigsty is compared with a horse stable, or the flight of a kite compared with a swallow or the song of a blackbird with the crowing of a cock. One goal of biodynamic farmers is to enable such diversity, which in turn shapes life on the farm. We can also observe how it influences the quality of the produce: wine growers report improvements to their wine when they succeed in creating a space for animals in their vineyards.

Mechthild Knösel's cattle herd

Mechthild Knösel is responsible for the cattle at Hofgut Rengoldshausen, near Lake Constance in Germany. She treats the animals with respect and ensures that they have good surroundings and live under species-appropriate conditions. The focus of Knösel's forty-strong herd are the milking cows – a pedigree Swiss Brown breed. By keeping this dual-purpose breed, it is possible to have both meat and milk production. The calves are raised with their mothers and remain fully integrated with the herd on the farm. The males are used for breeding or slaughtered later on the farm.

The feed used for the herd consists entirely of grass: fresh in the summer, then first or second cut hay in winter. The animals spend over eight months out at pasture. Knösel brings the young heifers into the milking parlour nine months before calving and when the time comes for them to be milked they enter without stress. Knösel's personal relationship to her animals extends to slaughtering them in a fear- and stress-free way. She kills the animals on the farm herself before having the meat prepared by the butcher and subsequently selling it from the farm.

In 2022, the Swiss Association for Biodynamic Agriculture decided that calves should remain for at least 120 days on the farm where they were born (or a biodynamic partner farm) to strengthen their immune system, and only then be transferred to a fattening or breeding unit. To allow the farms time to make the necessary adaptations, the new guidelines are being introduced in stages and will only come into full force by the end of 2030. This regulation comes in response to the discovery, both in research and in practice, that to become sufficiently stabilised, the calf's immune system needs this length of time to develop. There is otherwise the risk that antibiotics might have to be used after the transfer of the calves to another farm.

Appendix 1:
Summary of the Contents of
the Agriculture Course

The eight lectures of the Agriculture Course, given by Rudolf Steiner in June 1924, laid the foundations for biodynamics and made a lasting impact on the world of organic agriculture.

Lecture One (June 7, 1924)
Rudolf Steiner introduces the course by referring to the economic difficulties facing agriculture following the First World War and how a healthy agriculture is the basis for a healthy economy. He goes on to describe how plant growth is influenced not only by the soil and immediate environment, but even by the cosmos itself. Plants have a connection to their planetary surroundings: Saturn, Jupiter and Mars work via the silica substances on the nutritional quality of plants; the Moon, Venus and Mercury work via the calcium substances on the force of reproduction.

Lecture Two (June 10, 1924)
The concepts of the farm organism and farm individuality are introduced. Silica, lime and clay form the foundation of the

farm organism, with the animals providing the indispensable manure that builds up the fertility of the site through a closed cycle of substances. The farm organism lives within the polarity of above and below, sun and earth, cosmic and earthly. Compared to human beings, the farm individuality can be thought of as being upside down: the upper part, or head, is below ground; the lower part, or metabolic-limb system, is above in the light and the air; and the soil forms a kind of diaphragm.

Lecture Three (June 11, 1924)

Steiner discusses the five substances that make up protein: sulphur, carbon, oxygen, nitrogen and hydrogen. These elements are the material expression of active spiritual principles: sulphur is the bearer of the spiritual forces that make materialisation possible, carbon of the form-building forces, oxygen of life, nitrogen of the powers of sensation, and hydrogen has the task of leading from the material into the spiritual again. These five substances work together with calcium and silica. The working of spirit, soul and life elements are expressed in the activity of substances on the farm.

Lecture Four (June 12, 1924)

Steiner describes the working together of forces and substances using the example of nutrition: the food we eat is not meant to be deposited as substance, but instead provide the body with the forces it needs to be active and inwardly mobile. He discusses the role of compost and humus formation, before introducing the spray preparations:

horn manure and horn silica. Horn manure promotes a healthy soil and strong root growth; horn silica mediates light influences and enhances quality in the leaf-, flower- and fruit-forming processes.

Lecture Five (June 13, 1924)
Steiner introduces the biodynamic compost preparations: yarrow, chamomile, stinging nettle, oak bark, dandelion and valerian. He explains in each case how to make them. This involves surrounding the plant substances with animal organ material and allowing the seasonal influences and wider surroundings to work on them. The preparations are applied in homeopathic quantities to compost, manure or slurry. They have an organic and dynamic effect on the manure and fertilised soil.

Lecture Six (June 14, 1924)
This lecture deals with the issues of weeds, pests and plant diseases. The cosmic view described in the first two lectures is used to explain the powerful force of reproduction in annual plants. This comes from Venus, Mercury and especially the Moon. Burning the seeds of weeds and scattering the ashes over the fields can help prevent weed growth. A similar 'ashing' procedure is described for animal pests. To treat plant diseases the over-abundant forces of the Moon are deflected by spraying horsetail (*Equisetum arvense*) tea around the affected plants.

Lecture Seven (June 15, 1924)

This lecture is about the principles and practice of landscaping by agriculture. The lives of plants and animals depend on one another: birds and trees belong together, and there are corresponding connections between mammals and shrubs or broad leaf trees, insects and herbs, as well as between parasites and moist biotopes. Balancing areas of agricultural land with ecological reserves is encouraged. The farm's 'household of nature' is healthy when the 'giving' of plants and the 'taking' of animals is in balance.

Lecture Eight (June 16, 1924)

In this lecture Steiner deals with feeding. How can a species-appropriate feeding regime for farm animals be developed? Root fodder works primarily on the head in terms of nutrition, and it is from the head of the young animal that the rest of the organism is formed. That is why it is helpful to supplement the diet of calves with carrots. For milk-producing animals, leafy plants are recommended, especially legumes such as clover and lucerne. Flowers and seeds are fed to fattening animals to strengthen their muscular tissues. The course concludes by considering the closed cycle of substances in a farm organism, which helps develop the farm individuality.

Appendix 2:
A Timeline of Selected Events

1921: The Research Institute of the Natural Science Section is founded at the Goetheanum in Dornach, Switzerland.

1923: The first horn manure preparation is made following Rudolf Steiner's indications. Cow horns are buried not far from the Goetheanum and left in the ground over the winter. They are dug up the following spring.

1924: **June 7–16:** Rudolf Steiner gives the Agriculture Course in Koberwitz near Breslau (now Poland) for 130 participants. This provides the inspiration for the biodynamic movement. During the course, the Experimental Circle of the Anthroposophical Society is founded to carry out practical research into Steiner's ideas. It later became known as the Experimental Circle of Anthroposophical Farmers.

1925: **March 30:** Rudolf Steiner dies at the Goetheanum.

1928: Marienhöhe farm, to the east of Berlin, is purchased and converted to biodynamics. It continues to exist as a biodynamic farm to this day.
The Demeter trademark is registered as a symbol for biodynamic products for urban customers.

1938: Ehrenfried Pfeiffer's book *Soil Fertility, Renewal and Preservation* is published. Numbered copies of the Agriculture Course are still restricted to selected professionals and would remain so until the 1960s. Pfeiffer's book is therefore the first to make biodynamic agriculture available to the wider public.

1941: The Nazi Party bans all biodynamic organisations in Germany.

1954: The Demeter Federation is founded at Dottenfelderhof, Bad Vilbel, Germany.

1955: **January 1:** The Demeter Federation's nine-point constitution is created. It sets out rules for quality testing and trademark protection.

1962: Rachel Carson's book *Silent Spring* is published in the USA.

1971: Bioland founded in Germany

1973: Founding of FiBL, the Research Institute of Organic Agriculture in Switzerland

1977: FiBL begins the long-term DOK trials to compare three growing systems: biodynamic (D), organic (O) and conventional (K). The scientific study continues to this day.

1979: Sekem, a biodynamic community, is founded in the Egyptian desert by Ibrahim Abouleish. It is the start of the partnership between agricultural organisations, universities, research institutes and the German Demeter Federation to cultivate desert land for agricultural use.

1981: Bio Suisse is founded in Switzerland. It is the umbrella organisation for Swiss organic producers and the leading Swiss organic organisation.

1982: Naturland is founded in Germany.

1997: Demeter International is founded.

2020: Biodynamic Federation Demeter International (BFDI) is founded.

Notes

1. The Agriculture Course

1. See Bernard Jarman, *The Biodynamic Movement in Britain,* Chapter 1: The Beginning.
2. Keyserlingk, *The Birth of a New Agriculture.*
3. Steiner, Rudolf, *Agriculture Course: The Birth of the Biodynamic Method,* Rudolf Steiner Press, UK 2004, p. 5f. (This refers to the 1958 translation of the Agriculture Course by George Adams, which includes a preface by Ehrenfried Pfeiffer in which he describes the making of the horn preparations under Steiner's guidance. All other references to the Agriculture Course are from the edition published by Rudolf Steiner Press in 2024, translated by Simon Blaxland-De Lange.)
4. Steiner, *Agriculture,* p. 191.
5. Ibid., p. 194.
6. For more information see www.forschungsring.de.
7. Steiner, *Agriculture,* pp. 195f.

2. Key Concepts and Practices of the Agriculture Course

1. Steiner, *Agriculture,* p. 31.
2. Ibid., p. 35.
3. Ibid., p. 191.
4. Ibid., pp. 31f.
5. Ibid., p. 26.
6. Ibid., pp. 19f.
7. Ibid., p. 36.
8. Ibid., p. 46.
9. Klett, *The Foundations and Principles of Biodynamic Preparations,* p. 13.

10. Ibid., p. 35.
11. Ibid., p. 49.
12. Steiner, *Agriculture*, p. 93.
13. Ibid., pp. 136f.

3. The Historical Impact of the Agriculture Course

1. 'Albert Howard', *Wikipedia*, last updated: July 6, 2024, https://en.wikipedia.org/wiki/Albert_Howard
2. From the foreword to *An Agricultural Testament* by Sir Albert Howard.
3. For more information on the biodynamic movement during the Third Reich see 'Opportunism served "to save the soil"' by Sebastian Jüngel. Available at: https://www.sektion-landwirtschaft.org/en/news/sv/opportunism-served-to-save-the-soil
4. 'Hans Peter Rusch', *Wikipedia*, last updated: April 2, 2021, https://de.wikipedia.org/wiki/Hans_Peter_Rusch
5. 'Permaculture', *Wikipedia*, last updated: September 20, 2024, https://en.wikipedia.org/wiki/Permaculture
6. '*Tomorrow*' (2015 film), *Wikipedia*, last updated: September 4, 2024, https://en.wikipedia.org/wiki/Tomorrow_(2015_film)
7. Carson, *Silent Spring*, p. 130.
8. Matile, Philippe, 'Attackierte Lebensbasis: Grenzen der Kuntsdungerwirtschaft' [Attack on the Foundations of Life: The Limitations of Artificial Fertiliser Use in Agriculture], *Die Tat*, vol. 31, no. 249, p. 3.
9. Selawry, *Ehrenfried Pfeiffer*, p. 124.
10. For more information go to https://www.teikeicoffee.org/en/
11. Simpson, Emma, 'Dairy giant Arla warns of supply issues unless farmers paid more', *BBC News*, March 25, 2022, https://www.bbc.co.uk/news/business-60825516
12. Ortiz-Álvarez et al., 'Network Properties of Local Fungal Communities Reveal the Anthropogenic Disturbance Consequences of Farming Practices in Vineyard Soils', *ASM Journals*, vol. 6 no. 3, May 4, 2021, available at: https://journals.asm.org/doi/10.1128/msystems.00344-21
13. Soustre-Gacougnolle et al., 'Responses to Climatic and Pathogen Threats Differ in Biodynamic and Conventional Vines', *Scientific Reports*, November 15, 2018, available at: https://www.nature.com/articles/s41598-018-35305-7

14. See https://www.hs-geisenheim.de/research/departments/ general-and-organic-viticulture/department-of-general-organic-viticulture/?L=1
15. 'La dynamisation, le grand œvre des biodynamistes' [Dynamisation, the great work of biodynamicists], *La revue du vin de France*, no. 589.
16. Malnic, Evelyne, *La vigne, le vin et le bio* [Vines, wine and organic], p. 144.
17. See https://www.biodyvin.com
18. Steiner, *Agriculture*, p. 8.

4. The Biodynamic Movement 100 Years On

1. Steiner, *Agriculture*, p. 134.
2. Hurter, *Biodynamic Preparations Around the World*, pp. 221f.
3. For more information go to www.volkman.com.br/a-fazenda
4. https://prizeforcities.org/#id-155
5. For more information go to: www.wri.org/insights/rosario-urban-farming-tackles-climate-change
6. Lattuca, Atonio Luis, '"Agric-cultiver" les villes' ['Agri-cultivating' cities], *Biodynamis*, no. 116/2021, p. 46.
7. Ibid.
8. For more information go to: www.gramona.com/en/aliances-per-la-terra-work
9. See Hurter, *Biodynamic Preparations Around the World*, Chapter 15: Binta Shah, Supa Biotech (P) Ltd, India, pp. 311–32.
10. Shiva, Vandana, *Wer ernährt die Welt wirklich?* [Who Really Feeds the World?], p. 232.
11. For more information go to: www.navdanya.org
12. Rist, Michael, *Artgemässe Nutztierhaltung* [Species-appropriate Livestock Farming], p. 37.

Bibliography

Abouleish, Helmy and Arlt, Christine, *The Sekem Effect: How a Sustainable Community Can Transform Egypt and the World*, Floris Books, UK 2025.

Balfour, Lady Eve, *Living Soil*, Faber & Faber, UK 1975.

Carson, Rachel, *Silent Spring*, Penguin Books, UK 2000.

Groh, Trauger and McFadden, Steve, *Farms of Tomorrow: Community Supported Farms and Farms Supporting Communities*, Biodynamic Farming and Gardening Association, USA 1990.

—, *Farms of Tomorrow Revisited: Community Supported Farms and Farms Supporting Communities*, Biodynamic Farming and Gardening Association, USA 1997.

Howard, Albert, *An Agricultural Testament*, Benediction Classics, USA 2010.

—, *The Waste Products of Agriculture: Biodynamic and Organic Composting*, Oxford City Press, UK 2011.

Hurter, Ueli (ed.), *Biodynamic Preparations Around the World: Insightful Case Studies from Six Continents*, Floris Books, UK 2021.

Jarman, Bernard, *The Biodynamic Movement in Britain: A History of the First 100 Years*, Floris Books, UK 2024.

Keyserlingk, Adalbert Graf von, *The Birth of a New Agriculture: Koberwitz 1924 and the Introduction of Biodynamics*, Temple Lodge Publishing, UK 2009.

Klett, Manfred, *The Foundations and Principles of Biodynamic Preparations*, Floris Books, UK 2023.

Malnic, Evelyne, *La vigne, le vin et le bio: L'avenir de la viticulture s'écrit en bio-logique et dynamique* [Vines, Wine and Organic: The Future of Viticulture is Written in Organic and Dyamic], France Agricole, France 2021.

Meyer, Thomas, *Ehrenfried Pfeiffer: A Modern Quest for the Spirit*, Mercury Press, UK 2010.

Northbourne, Christopher James (Lord), *Look to the Land*, Sophia Perennis, USA 2003.

Paull, John, 'The Betteshanger Summer School: Missing Link between Biodynamic and Organic Farming', ResearchGate, 2014. Available at: https://www.researchgate.net/publication/228440 388_The_Betteshanger_Summer_School_Missing_Link_ between_Biodynamic_and_Organic_Farming.

—, 'The Rachel Carson Letters and the Making of Silent Spring', in *SAGE Open*, July-September 2013. Available at: https://journals. sagepub.com/doi/full/10.1177/2158244013494861

Pfeiffer, Ehrenfried, *Biodynamic Farming and Gardening: Renewal and Preservation of Soil Fertility*, SteinerBooks, USA 2021.

Rist, Michael, *Artgemässe Nutztierhaltung. Ein Schritt zum wesensgemäßen Umgang mit der Natur* [Species-appropriate Livestock Farming: A Step Towards Dealing with Nature Appropriately], Freies Geistesleben, Germany 1989.

Selawry, Alla, *Ehrenfried Pfeiffer: Pionier Spiritueller Forschung und Praxis* [Ehrenfried Pfeiffer: Pioneer of Spiritual Research and Practice], Goetheanum Verlag, Switzerland, 1987.

Shiva, Vandana, *Wer ernährt die Welt wirklich? Das Versagen der Agrarindustrie und die notwendige Wende zur Agrarökologie* [Who Really Feeds the World?: The Failure of the Agriculture Industry and the Necessary Shift to Agroecology], Neue Erde Verlag, Germany 2021.

Steiner, Rudolf, *Agriculture: Spiritual-Scientific Foundations for Agricultural Renewal* (CW327), Rudolf Steiner Press, UK 2024.

—, *Towards Social Renewal: Rethinking the Basis of Society* (CW23), Rudolf Steiner Press, UK 2009.

—, *Rethinking Economics: Lectures and Seminars on World Economics* (CW340 & 341), Rudolf Steiner Press, UK 2013.

Picture Credits

Public domain:
Figures 1.1, 3.2, 3.4, 3.6, 4.3 and 4.6.

Captioned illustrations:
Bundesarchiv: Figures 3.3 and 3.11.
Demeter: Figure 4.1.
Dottenfelderhof: Figure 4.7.
FiBL (www.fibl.org): Figures 3.9 and 3.10.
Fischer, Charlotte: Figure 1.3.
Floris Books: Figure 3.5.
Gramona (www.gramona.com): Figure 4.5.
Grubenmann, Myriam: Figure 3.8.
Marienhöhe (www.hofmarienhoehe.de): Figures 3.1a and 3.1b.
Moriconi, Silvio: Figure 4.4.
Onneken, Johannes: Figure 3.7.
Rudolf Steiner Archive: Figures 1.2, 2.1, 2.2.
Sekem (www.sekem.com): Figure 4.2.

Uncaptioned illustrations:
Fischer, Charlotte: pp. 65, 74, 79, 83, 108 and 111.
Kunz, Peter (GZPK): p. 104.
Lieberherr, Cristina: p. 8.
Timbuktu Collective: p. 107.
Wahl, Verena: pp. 67 and 101.

Index

Floris Books

For news on all our **latest books**,
and to receive **exclusive discounts**,
join our mailing list at:

florisbooks.co.uk

Plus subscribers get a FREE book
with every online order!

We will never pass your details to anyone else.

www.ingramcontent.com/pod-product-compliance
Lightning Source LLC
Chambersburg PA
CBHW051259020426
42333CB00026B/3270